SpringerBriefs on PDEs and Data Science

Editor-in-Chief

Enrique Zuazua, Department of Mathematics, University of Erlangen-Nuremberg, Erlangen, Bayern, Germany

Series Editors

Irene Fonseca, Department of Mathematical Sciences, Carnegie Mellon University, Pittsburgh, PA, USA

Franca Hoffmann, Hausdorff Center for Mathematics, University of Bonn, Bonn, Germany

Shi Jin, Institute of Natural Sciences, Shanghai Jiao Tong University, Shanghai, Shanghai, China

Juan J. Manfredi, Department of Mathematics, University Pittsburgh, Pittsburgh, PA, USA

Emmanuel Trélat, CNRS, Laboratoire Jacques-Louis Lions, Sorbonne University, PARIS CEDEX 05, Paris, France

Xu Zhang, School of Mathematics, Sichuan University, Chengdu, Sichuan, China

SpringerBriefs on PDEs and Data Science targets contributions that will impact the understanding of partial differential equations (PDEs), and the emerging research of the mathematical treatment of data science.

The series will accept high-quality original research and survey manuscripts covering a broad range of topics including analytical methods for PDEs, numerical and algorithmic developments, control, optimization, calculus of variations, optimal design, data driven modelling, and machine learning. Submissions addressing relevant contemporary applications such as industrial processes, signal and image processing, mathematical biology, materials science, and computer vision will also be considered.

The series is the continuation of a former editorial cooperation with BCAM, which resulted in the publication of 28 titles as listed here: https://www.springer.com/gp/mathematics/bcam-springerbriefs

Roberto Alicandro • Nadia Ansini •
Andrea Braides • Andrey Piatnitski •
Antonio Tribuzio

A Variational Theory of Convolution-Type Functionals

 Springer

Roberto Alicandro
Department of Electrical and Information
Engineering
University of Cassino and Southern Lazio
Cassino, Frosinone, Italy

Nadia Ansini
Department of Mathematics
Sapienza University of Rome
Rome, Italy

Andrea Braides
SISSA
Trieste, Italy

Andrey Piatnitski
Department of Technology
UiT The Arctic University of Norway
Narvik, Norway

Antonio Tribuzio
Institute for Applied Mathematics
University of Heidelberg
Heidelberg, Baden-Württemberg, Germany

ISSN 2731-7595   ISSN 2731-7609 (electronic)
SpringerBriefs on PDEs and Data Science
ISBN 978-981-99-0684-0   ISBN 978-981-99-0685-7 (eBook)
https://doi.org/10.1007/978-981-99-0685-7

Mathematics Subject Classification: 49J45, 49J55, 74Q05, 35B27, 35B40, 45E10

This Springer imprint is published by the registered company Springer Nature Singapore Pte Ltd.
The registered company address is: 152 Beach Road, #21-01/04 Gateway East, Singapore 189721, Singapore

Preface

These notes stem from a group activity in Rome stimulated by a visit of Andrey Piatnitski at the University of Rome Tor Vergata, where he was also giving a course on the homogenization of convolution operators. Noting the timely nature of that subject and the potentiality of applications in different directions, from evolution phenomena with long-range diffusion to Data Science, the working group focussed on providing an abstract framework for a comprehensive theory of convolution-type functionals. In that project, we had in mind on one hand the analogy with theories regarding the passage discrete-to-continuum developed in the last 20 years, and on the other hand a number of previous homogenization results for convolution operators both in the static and dynamic framework. The result was a quite complete formalization based on a very general compactness theorem in the spirit of the De Giorgi localization methods for Γ-convergence, and a series of preliminary abstract results, such as embeddings, Poincarè inequalities, etc., with a number of applications, in particular to stochastic homogenization, to energies on point clouds and to gradient flows, which are just some of the potential directions of the theory. The results presented here are intended to provide an environment, together with the related technical tools, in which to frame both static and dynamic problems related to multiple-scale variational models where non-local interactions at a small scale are involved. As such, the notes are addressed to a readership of Mathematical Analysts and Applied Mathematicians. Beside the material elaborated in the working group, the present notes are complemented by references to peridynamics, in whose terminology our analysis gives a general framework for the limit as the horizon tends to zero, and a revisitation of some recent works on perforated domains in the light of the compactness theorem.

We are very grateful to Enrique Zuazua for having proposed us to contribute to this book series. We also acknowledge the fruitful interaction with Valeria Chiadò Piat, Lorenza D'Elia, Carolin Kreisbeck and Elena Zhizhina on issues connected with the content of these notes.

The authors acknowledge the support of the Italian MIUR Excellence Department Project awarded to the Department of Mathematics, University of Rome Tor Vergata, CUP E83C18000100006. Antonio Tribuzio was partially supported by the Deutsche Forschungsgemeinschaft through SPP 2256, project ID 441068247.

Cassino, Italy   Roberto Alicandro
Rome, Italy   Nadia Ansini
Trieste, Italy   Andrea Braides
Narvik, Norway   Andrey Piatnitski
Heidelberg, Germany   Antonio Tribuzio
October 2022

Contents

Chapter 1
Introduction

Abstract We provide a general treatment of a class of functionals modeled on convolution energies with kernel having finite p-moments. Such model energies approximate the p-th norm of the gradient as the kernel is scaled by letting a small parameter ε tend to 0. We first provide the necessary functional-analytic tools to show coerciveness of families of such functionals with respect to strong L^p convergence. The main result is a compactness and integral-representation theorem which shows that limits of convolution-type energies are local integral functionals with p-growth defined on a Sobolev space. This result is applied to obtain periodic homogenization results, to study applications to functionals defined on point-clouds, to stochastic homogenization and to the study of limits of the related gradient flows.

Keywords Nonlocal energies · Peridynamics · Population dynamics · Compactness · Integral representation · Periodic homogenization · Stochastic homogenization · Point clouds · Gradient flows

Scope of these notes is a general asymptotic analysis of families of non-local functionals of the form

$$\int_{\Omega\times\Omega} W_\varepsilon(x, y, u(y) - u(x))\, dx\, dy, \qquad (1.1)$$

with Ω a Lipschitz domain in $\mathbb{R}^d$, as the parameter ε tends to 0, and the energies 'concentrate' on the diagonal $x = y$. Energies of this form are customary in the theory of *peridynamics* (see e.g. [9, 22, 23, 36–39]), where ε represents the range of interaction between points, so that indeed the energies above are of the form

$$\int_{\{(x,y)\in\Omega\times\Omega:|x-y|\leq\varepsilon\}} W_\varepsilon(x, y, u(y) - u(x))\, dx\, dy. \qquad (1.2)$$

© The Author(s), under exclusive license to Springer Nature Singapore Pte Ltd. 2023   1
R. Alicandro et al., *A Variational Theory of Convolution-Type Functionals*,
SpringerBriefs on PDEs and Data Science,
https://doi.org/10.1007/978-981-99-0685-7_1

A prototypical assumption on W_ε is that

$$W_\varepsilon(x, y, z) = \frac{1}{\varepsilon^d} W\left(x, \frac{y-x}{\varepsilon}, \frac{z}{\varepsilon}\right). \tag{1.3}$$

If $u \in C^1(\Omega; \mathbb{R}^m)$ then we may approximate $u(x) - u(y)$ with $\nabla u(x)(x - y)$, so that, after a change of variables $y = x + \varepsilon\xi$, formally we can write the energies above approximatively as

$$\int_{\Omega \times B_1(0)} W\left(x, \xi, \frac{u(x + \varepsilon\xi) - u(x)}{\varepsilon}\right) dx d\xi \sim \int_{\Omega \times B_1(0)} W\left(x, \xi, \nabla u(x)\xi\right) dx d\xi. \tag{1.4}$$

This formula suggests that the limit may be a local functional representable as an integral. Under some growth and convexity hypothesis on W, indeed it has been proved that the Γ-limit is defined on some Sobolev space $W^{1,p}(\Omega; \mathbb{R}^m)$ and has the local form

$$F(u) = \int_\Omega W_0(x, \nabla u)\, dx, \quad \text{where } W_0(x, M) = \int_{B_1(0)} W\left(x, \xi, M\xi\right) d\xi \tag{1.5}$$

for every $m \times d$ matrix M (see [5, 6, 31]). This argument will be made rigorous as a very particular case of the results in the following.

In order to build up a general theory for functionals (1.1), we start with considering *convolution functionals*; i.e., functionals of the form

$$\frac{1}{\varepsilon^{d+p}} \int_{\Omega \times \Omega} a\left(\frac{y-x}{\varepsilon}\right) |u(y) - u(x)|^p dx\, dy, \tag{1.6}$$

as a simpler class of prototype energies. In formula (1.6) Ω is a Lipschitz domain in $\mathbb{R}^d$ and a is a sufficiently integrable positive kernel; namely, a satisfies

$$\int_{\mathbb{R}^d} a(\xi)|\xi|^p\, d\xi < +\infty. \tag{1.7}$$

This choice corresponds to choosing $W(x, \xi, z) = a(\xi)|z|^p$ in (1.3). Functionals as in (1.6) can be seen as an approximation and a generalization of an L^p-norm of the gradient of u, and as such have been used e.g. in non-local approaches to phase-transition problem (see Alberti and Bellettini [1]). Again, if u is of class $C^1(\Omega)$, we may approximate $u(y) - u(x)$ with $\langle \nabla u(x), y - x \rangle$, so that, up to an error which can be neglected as $\varepsilon \to 0$, energies (1.6) can be rewritten as

$$\frac{1}{\varepsilon^d} \int_{\Omega \times \Omega} a\left(\frac{y-x}{\varepsilon}\right) \left|\left\langle \nabla u(x), \frac{y-x}{\varepsilon}\right\rangle\right|^p dx\, dy. \tag{1.8}$$

After the change of variables $y = x + \varepsilon\xi$ and letting $\varepsilon \to 0$, we obtain

$$\int_\Omega \|\nabla u\|_a^p \, dx, \text{ where } \|z\|_a^p = \int_{\mathbb{R}^d} a(\xi)|\langle z, \xi\rangle|^p d\xi. \tag{1.9}$$

In the particular case when a is radially symmetric we obtain a constant times the p-th power.

Limits of energies similar to (1.6), of the form

$$\frac{1}{\varepsilon^d} \int_{\Omega \times \Omega} a\left(\frac{y-x}{\varepsilon}\right) \frac{|u(y) - u(x)|^p}{|y-x|^p} dy \, dx, \tag{1.10}$$

have also been studied by Bourgain et al. [10] as an alternative definition of the L^p-norm of the gradient of a Sobolev function (see e.g. also [35]), as part of a number of works stemming from a general interest towards non-local variational problems arisen in the last twenty years (see e.g. the survey paper [21] or the book [18]). It is worth recalling that a non-linear version of functionals (1.6) with truncated quadratic potentials had been previously proposed as an approximation of the Mumford-Shah energy by De Giorgi and subsequently studied by Gobbino [26] (see also [27], and corresponding energies in peridynamics in [30]). Furthermore, discrete energies of the form

$$\frac{1}{\varepsilon^{d+p}} \sum_{i,j \in \mathcal{L}} a_{ij} |u_j - u_i|^p, \tag{1.11}$$

where $\mathcal{L}$ is a d-dimensional lattice, $u : \varepsilon\mathcal{L} \to \mathbb{R}^m$ and $u_i = u(\varepsilon i)$, have been widely investigated as a discrete approximation of integral functionals of p-growth (see e.g. [2, 11, 12, 33]; see also [4, 16] for static and dynamic lattice problems with interfaces). Such energies can be seen as a discrete version of functionals (1.6).

In the case $p = 2$, some energies of the form (1.6) derive from models in population dynamics where macroscopic properties can be reduced to studying the evolution of the first-correlation functions describing the population density u in the system [24, 28]. With that interpretation in mind, in the simplest formulation one can consider their perturbations

$$\frac{1}{\varepsilon^{d+2}} \int_{\Omega \times \Omega} b_\varepsilon(x, y) a\left(\frac{y-x}{\varepsilon}\right) |u(y) - u(x)|^2 dx \, dy, \tag{1.12}$$

that may take into account inhomogeneities of the environment encoded in the (non-negative) coefficient b_ε. Functionals modeled on these energies, with b_ε obtained by scaling a given periodic function b or with a random coefficient $b_\varepsilon = b_\varepsilon^\omega$ depending on the realization of a random variable, have been considered in the context of homogenization for u scalar in [14, 15], producing a limit elliptic homogeneous functional

$$\int_{\Omega} \langle A_{\mathrm{hom}} \nabla u(x), \nabla u(x) \rangle \, dx \,. \tag{1.13}$$

This limit can be expressed in terms of Γ-convergence and implies the convergence of related minimum problems. Another type of perturbations of functionals (1.6) is encountered in a different application to Data Science, studied by García Trillos and Slepcev [25], who examine energies approximating the total variation of u of the form (1.6)

$$\frac{1}{\varepsilon^{d+p}} \int_{\Omega \times \Omega} a \left(\frac{T_{\varepsilon}(y) - T_{\varepsilon}(x)}{\varepsilon} \right) |u(y) - u(x)|^p dx \, dy, \tag{1.14}$$

when $p = 1$, with $T_{\varepsilon} : \Omega \to \Omega$ and discuss its stability in terms of the convergence of T_{ε} to the identity. This result has been extended to the case of energies with p-growth for $p > 1$ in [20] (see also [19] in the context of free-discontinuity problems).

In this book, we provide a general treatment of 'convolution-type' energies (1.1) modeled on (1.6) and (1.12) (and including also the case (1.14) with $p > 1$) that allow for a general non-linear dependence on $u(x) - u(y)$ and inhomogeneity on x and y. More precisely, the functionals that we will consider are of the form

$$\frac{1}{\varepsilon^{d+p}} \int_{\Omega \times \Omega} f_{\varepsilon}(x, y, u(x) - u(y)) \, dx \, dy, \tag{1.15}$$

with $p > 1$; i.e., we consider the scaled function $f_{\varepsilon} = \varepsilon^{d+p} W_{\varepsilon}$ in (1.1). The functions $f_{\varepsilon} : \Omega \times \Omega \times \mathbb{R}^m \to \mathbb{R}$ are quite general. In order to compare functionals (1.15) with the energies defined in (1.6) we assume that Ω is a bounded domain and that there exist two kernels a_1 and a_2 such that

$$a_1 \left(\frac{y - x}{\varepsilon} \right) (|z|^p - \varepsilon^p) \le f_{\varepsilon}(x, y, z) \le a_2 \left(\frac{y - x}{\varepsilon} \right) (|z|^p + \varepsilon^p). \tag{1.16}$$

A non-degeneracy condition for the limit is ensured e.g. by assuming that

$$a_1(\xi) \ge c_0 \quad \text{if } |\xi| \le r_0$$

for some $c_0, r_0 > 0$, while a decay condition in a_2 provides that the limit be finite exactly on $W^{1,p}(\Omega; \mathbb{R}^m)$. Note that, for a wider applicability of this analysis, considering a dependence on ε of the kernel $a_2 = a_2^{\varepsilon}$ is also necessary. Since this general form of the kernels may cause the limit energy to be non-local, some uniform conditions on the decay of the a_2^{ε} must be required to ensure that the limit be a local integral energy. These assumptions will be stated precisely in Sect. 2.2. The

central result of this book is that, up to subsequences, the energies above converge
to an energy of the form

$$\int_\Omega f_0(x, \nabla u)\,dx, \tag{1.17}$$

with domain $W^{1,p}(\Omega; \mathbb{R}^m)$. This convergence is expressed as a Γ-limit with respect
to the L^p-topology in Ω. This is justified by a Compactness Theorem, which
states that, if a is the characteristic function of a ball, any sequence $\{u_\varepsilon\}$ bounded
in $L^p(\Omega)$ with equibounded energies (1.6) admits a subsequence converging to
some $u \in W^{1,p}(\Omega; \mathbb{R}^m)$ with respect to the $L^p(\Omega; \mathbb{R}^m)$-topology. This result is
complemented by the validity of suitable Poincaré inequalities, which allow to
prove the equi-coerciveness of the functionals subjected to boundary data and the
application of the direct methods of Γ-convergence to the asymptotic description
of minimum problems. It is worth noting that even in the 'homogeneous' case of
functionals of the form (1.1) with W_ε as in (1.3) and $W = W(\xi, z)$ not convex in
z the functionals in (1.1) are not lower semicontinuous with respect to the weak
L^p-convergence, and their lower-semicontinuous envelope is not even an integral
functional [29, 32]. In particular, it is then not possible to justify a simple expansion
argument as in (1.4).

We also include various applications. First, to the homogenization of non-local
functionals of the form

$$\frac{1}{\varepsilon^{d+p}} \int_{\Omega \times \Omega} f\left(\frac{x}{\varepsilon}, \frac{y}{\varepsilon}, u(x) - u(y)\right) dx\,dy, \tag{1.18}$$

with f periodic in the first variable. In this case the limit integrand f_0 in (1.17) is
independent of x and can be characterized by a non-local asymptotic formula, which
can be further simplified to a non-local cell-problem formula if f is convex in the
last variable. A second application is to a class of non-local functionals of the form

$$\frac{1}{\varepsilon^{d+p}} \int_{\Omega \times \Omega} f_\varepsilon(T_\varepsilon(x), T_\varepsilon(y), u(y) - u(x))\rho(x)\rho(y)\,dx\,dy, \tag{1.19}$$

which generalize (1.14). If the image of T_ε is discrete these energies can be
interpreted as a continuum interpolation of discrete energies. In particular, following
[25] we can use these functionals to describe the behavior of energies defined
on point clouds. A third application is a stochastic homogenization theorem; i.e.,
the characterization of the limit of functionals in (1.18) when the integrand $f =
f(\omega)$ is a statistically homogeneous (in the first variable) random function defined
through a measure-preserving ergodic dynamical system. We characterize the limit
using an asymptotic non-local homogenization formula, and prove that the limit is
deterministic under ergodicity assumptions. Related results in a discrete setting can
be found e.g. in [3, 7, 8, 13]. Finally, we also treat some evolutionary problems
using the methods of minimizing movements if the functions f_ε are convex in the

last variable. In particular, we consider the homogenization case (1.18), and show
that the solutions of gradient flows for those energies, which take the form

$$\partial_t u_\varepsilon(t,x) = -\frac{1}{\varepsilon^{d+1}} \int_\Omega \nabla_z f\left(\frac{y}{\varepsilon}, \frac{x}{\varepsilon}, \frac{u_\varepsilon(t,x) - u_\varepsilon(t,y)}{\varepsilon}\right) dy$$
$$+ \frac{1}{\varepsilon^{d+1}} \int_\Omega \nabla_z f\left(\frac{x}{\varepsilon}, \frac{y}{\varepsilon}, \frac{u_\varepsilon(t,y) - u_\varepsilon(t,x)}{\varepsilon}\right) dy,$$

converge to the solution of the corresponding gradient flow for the limit homogenized energy. Previously, homogenization problems for parabolic equations with
linear periodic not necessarily symmetric convolution type operators have been
studied in [34], where homogenization results were obtained by means of two-scale
expansions technique.

The plan of the book is as follows. In Chap. 2 we introduce the necessary
notation and introduce the class of convolution-type energies under examination.
In particular, we change the formal appearance of our energies so as to highlight
the range of the interactions. We compare our hypotheses with the corresponding
ones for integral functionals, commenting on analogies and differences. We state
a weaker version of the coerciveness assumption that may be of use when dealing
e.g. with perforated domains (see [17]). We also introduce the special convolution
energies $G_\varepsilon[a]$ of type (1.6) and in particular G_ε^r when $a = \chi_{B_r}$, which are used
as comparison energies throughout the notes (in particular, since they are a lower
bound for the energies we consider, it is sufficient to state compactness results for
families of functions $\{u_\varepsilon\}$ with $G_\varepsilon^r(u_\varepsilon)$ bounded). Chapter 4 contains some general
results, that mirror the analog results in Sobolev spaces, of extension from Lipschitz
sets, compactness with respect the L^p convergence and Poincaré inequalities, where
the role of the p-th norm of the gradient is played by G_ε^r as $\varepsilon \to 0$. The fundamental
Lemma 4.1 allows to control long-range interactions with short-range interactions,
while Compactness Theorem 4.2 guarantees both a compact embedding in L^p for
sequences with equibounded G_ε^r-energies, and that their limits belong to $W^{1,p}$.
In Chap. 3 we consider the particular case of the limit of energies $G_\varepsilon[a_\varepsilon]$ with
varying a_ε, characterizing their limits. In particular, when $a_\varepsilon = a$ we obtain the Γ-
convergence to energy (1.9). This limit is used to provide lower and upper bounds
for the general case. The main result of the book is contained in Chap. 5, where
the general compactness and integral-representation Theorem 5.1 is proved using a
variation of the localization method of Γ-convergence, which is possible since, even
though convolution-type functionals are non-local, their non-locality 'vanishes' as
$\varepsilon \to 0$. An important technical result formalizing this observation is Lemma 5.1,
which states, in terms of functionals (1.6), that it is not restrictive to deal with
kernels a with bounded support, up to a truncation argument. Section 5.5 deals
with the convergence of minimum problems with Dirichlet boundary conditions.
Note that for convolution-type functionals such conditions must be imposed on
a neighbourhood of size of order ε of the boundary. Chapter 6 specializes the
description of the Γ-limit in the case of periodically oscillating energies, using

the integral-representation result, the truncation argument and the convergence of minimum problems to obtain homogenization formulas for the energy function of the Γ-limit, both of asymptotic type (Theorem 6.1 and formula (6.7)) taking into account interactions within cubes of diverging size and on periodic functions in the convex case (Theorem 6.2 and formula (6.18)). Note that in the latter we take into account interactions $u(x) - u(y)$ with x in the periodicity set and y in the whole space. In Chap. 7 we consider functionals of the form (1.19) and prove their equivalence to the corresponding functionals (1.15) when T_ε approaches the identity up to an error of order $o(\varepsilon)$, under some technical conditions on f_ε. This result is then applied to the analysis of energies defined on point clouds in Chap. 7.2. Chapter 8 contains a homogenization theorem in the stochastic setting (Theorem 8.2), whose proof generalizes the arguments utilized for the deterministic homogenization result, using a subadditive ergodic theorem by Krengel and Pyke (Theorem 8.1) to characterize an asymptotic homogenization formula. Finally, in Chap. 9 we study the convergence of the gradient flows associated to our energies in the convex case. A general approach by minimizing movements allows to deduce in particular the convergence of gradient flows in the case of the homogenization to the gradient flow of the homogenized limit, which is a standard parabolic equation (Theorem 9.3).

References

1. Alberti, G., Bellettini, G.: A non-local anisotropic model for phase transitions: asymptotic behaviour of rescaled energies. Eur. J. Appl. Math. **9**, 261–284 (1998)
2. Alicandro, R., Cicalese, M.: A general integral representation result for continuum limits of discrete energies with superlinear growth. SIAM J. Math. Anal. **36**, 1–37 (2004)
3. Alicandro, R., Cicalese, M., Gloria, A.: Integral representation results for energies defined on stochastic lattices and application to nonlinear elasticity. Arch. Ration. Mech. Anal. **200**, 881–943 (2011)
4. Alicandro, R., Braides, A., Cicalese, M., Solci, M.: Discrete Variational Problems with Interfaces. Cambridge University Press, Cambridge (2023)
5. Bellido, J.C., Mora-Corral, C., Pedregal, P.: Hyperelasticity as a Γ-limit of peridynamics when the horizon goes to zero. Calc. Var. Partial Differ. Equ **54**, 1643–1670 (2015)
6. Bellido, J.C., Cueto, J., Mora-Corral, C.: Γ-convergence of polyconvex functionals involving s-fractional gradients to their local counterparts. Calc. Var. Partial Differ. Equ. **60**(7) (2021)
7. Blanc, X., Le Bris, C., Lions, P.-L.: The energy of some microscopic stochastic lattices. Arch. Ration. Mech. Anal. **184**, 303–339 (2007)
8. Blanc, X., Le Bris, C., Lions, P.-L.: Stochastic homogenization and random lattices. J. Math. Pures Appl. (9) **88**, 34–63 (2007)
9. Bobaru, F., Foster, J.T., Geubelle, P.H., Silling, S.A.: Handbook of Peridynamic Modeling. Advances in Applied Mathematics. CRC press, Boca Raton (2016)
10. Bourgain, J., Brezis, H., Mironescu, P.: Another look at Sobolev spaces. In: Optimal Control and Partial Differential Equations, pp. 439–455. IOS Press, Amsterdam (2001)
11. Braides, A.: Discrete-to-continuum variational methods for lattice systems. In: Proceedings of the International Congress of Mathematicians–Seoul, vol. IV, pp. 997–1015. Kyung Moon Sa, Seoul (2014)
12. Braides, A., Kreutz, L.: An integral-representation result for continuum limits of discrete energies with multibody interactions. SIAM J. Math. Anal. **50**, 1485–1520 (2018)

13. Braides, A., Piatnitski, A.: Homogenization of surface and length energies for spin systems. J. Funct. Anal. **264**, 1296–1328 (2013)
14. Braides, A., Piatnitski, A.: Homogenization of random convolution energies. J. Lond. Math Soc. (2) **104**, 295–319 (2021)
15. Braides, A., Piatnitski, A.: Homogenization of convolution energies in periodically perforated domains. Adv. Calc. Var. **15**, 351–368 (2022)
16. Braides, A., Solci, M.: Geometric Flows on Planar Lattices. Pathways in Mathematics. Birkhäuser/Springer, Cham (2021)
17. Braides, A., Chiadò Piat, V., D'Elia, L.: An extension theorem from connected sets and homogenization of non-local functionals. Nonlinear Anal. **208**, 112316 (2021)
18. Bucur, C., Valdinoci, E.: Nonlocal Diffusion and Applications. Lecture Notes of the Unione Matematica Italiana, vol. 20. Springer, Cham (2016)
19. Caroccia, M., Chambolle, A., Slepčev, D.: Mumford-shah functionals on graphs and their asymptotics. Nonlinearity **33**, 3846–3888 (2020)
20. Crook, O.M., Hurst, T., Schönlieb, C.-B., Thorpe, M., Zygalakis, K.C.: PDE-inspired algorithms for semi-supervised learning on point clouds. arXiv preprint, arXiv:1909.10221v1
21. Di Nezza, E., Palatucci, G., Valdinoci, E.: Hitchhiker's guide to the fractional Sobolev spaces. Bull. Sci. Math. **136**, 521–573 (2012)
22. Diehl, P., Lipton, R., Wick, T., Tyagi, M.: A comparative review of peridynamics and phase-field models for engineering fracture mechanics. Comput. Mech. **69**, 1259–1293 (2022)
23. Du, Q., Zhou, K.: Mathematical analysis for the peridynamic nonlocal continuum theory. ESAIM: Math. Model. Numer. Anal. **45**, 217–234 (2011)
24. Finkelshtein, D., Kondratiev, Y., Kutoviy, O.: Semigroup approach to birth-and-death stochastic dynamics in continuum. J. Funct. Anal. **262**, 1274–1308 (2012)
25. García Trillos, N., Slepčev, D.: Continuum limit of total variation on point clouds. Arch. Ration. Mech. Anal. **220**, 193–241 (2016)
26. Gobbino, M.: Finite difference approximation of the Mumford-Shah functional. Commun. Pure Appl. Math. **51**, 197–228 (1998)
27. Gobbino, M., Mora, M.G.: Finite-difference approximation of free-discontinuity problems. Proc. R. Soc. Edinb. Sect. A **131**, 567–595 (2001)
28. Kondratiev, Y., Kutoviy, O., Pirogov, S.: Correlation functions and invariant measures in continuous contact model. Infin. Dimens. Anal. Quantum Probab. Relat. Top. **11**, 231–258 (2008)
29. Kreisbeck, C., Zappale, E.: Loss of double-integral character during relaxation. SIAM J. Math. Anal. **53**, 351–385 (2021)
30. Lipton, R.: Dynamic brittle fracture as a small horizon limit of peridynamics. J. Elast. **117**, 21–50 (2014)
31. Mengesha, T., Du, Q.: On the variational limit of a class of nonlocal functionals related to peridynamics. Nonlinearity **28**, 3999–4035 (2015)
32. Mora-Corral, C., Tellini, A.: Relaxation of a scalar nonlocal variational problem with a double-well potential. Calc. Var. Partial Differ. Equ. **59**(67) (2020)
33. Piatnitski, A., Remy, E.: Homogenization of elliptic difference operators. SIAM J. Math. Anal. **33**, 53–83 (2001)
34. Piatnitski, A., Zhizhina, E.A.: Homogenization of biased convolution type operators. Asymptot. Anal. **115**, 241–262 (2019)
35. Ponce, A.C.: A new approach to Sobolev spaces and connections to Γ-convergence. Calc. Var. Partial Differ. Equ. **19**, 229–255 (2004)
36. Silling, S.A.: Reformulation of elasticity theory for discontinuities and long-range forces. J. Mech. Phys. Solids **48**, 175–209 (2000)
37. Silling, S.A., Lehoucq, R.B.: Convergence of peridynamics to classical elasticity theory. J. Elast. **93**, 13–37 (2008)
38. Silling, S.A., Lehoucq, R.B.: Peridynamic theory of solid mechanics. Adv. Appl. Mech. **44**, 73–168 (2010)
39. Silling, S.A., Epton, M.A., Weckner, O., Xu, J., Askari, E.: Peridynamic states and constitutive modeling. J. Elast. **88**, 151–184 (2007)

Chapter 2
Convolution-Type Energies

Abstract In this chapter we formalize the assumptions on our families of convolution-type functionals. Such assumptions are stated in terms of some growth and integrability conditions. We explain and comment these hypotheses comparing them with the corresponding assumptions for families of local integral functionals commonly used in the literature.

Keywords Nonlocal functionals · Convolution kernels · Growth conditions · Non-degeneracy · Kernels of polynomial decay · Sign-changing kernels

2.1 Notation

In these notes $d, m \in \mathbb{N}$ will be fixed natural numbers denoting the dimension of the reference and target spaces of the functions we consider, respectively, and $p > 1$ will be a growth exponent. We let $\Omega \subset \mathbb{R}^d$ be a bounded open set with Lipschitz boundary. Note that for many results this regularity assumption on Ω may be removed up to considering local arguments.

For a generic point $x \in \mathbb{R}^d$ we write $x = (x_1, \ldots, x_d)$, we let $\{e_j\}_{j=1}^d$ be the *canonical basis* of $\mathbb{R}^d$. We write $\mathbb{S}^{d-1} = \{x \in \mathbb{R}^d : |x| = 1\}$. We let $\lfloor t \rfloor$ and $\lceil t \rceil$ denote the lower and upper integer part of $t \in \mathbb{R}$, respectively. Given $x \in \mathbb{R}^d$, we let $\lfloor x \rfloor$ denote the vector in $\mathbb{Z}^d$ whose components are the integer parts of the components of x; that is, $\lfloor x \rfloor = (\lfloor x_1 \rfloor, \ldots, \lfloor x_d \rfloor)$. Analogously $\lceil x \rceil = (\lceil x_1 \rceil, \ldots, \lceil x_d \rceil)$. If $x, y \in \mathbb{R}^d$ then $|x|$ denotes the norm of x and $\langle x, y \rangle$ the scalar product between x and y. $\mathbb{R}^{m \times d}$ denotes the space of $m \times d$ matrices with real entries; if $M \in \mathbb{R}^{m \times d}$ and $x \in \mathbb{R}^d$ then $Mx \in \mathbb{R}^m$ is defined by the usual row-by-column product. We use $SO(d) \subset \mathbb{R}^{d \times d}$ to denote the *group of rotations* of $\mathbb{R}^d$.

We let $B_r(x)$ (if $x = 0$, simply B_r) be the open ball of centre x and radius r. If A and B are subsets of $\mathbb{R}^d$ then $\mathrm{dist}(x, A) = \inf\{|z - x| : z \in A\}$ and $\mathrm{dist}(A, B) = \inf\{|z - x| : z \in A, x \in B\}$ denote the distance of x from A and from A to B, respectively. $\mathcal{A}(\Omega)$ will be the family of all open subsets of Ω and $\mathcal{A}^{\mathrm{reg}}(\Omega)$ the subfamily of open subsets with Lipschitz boundary. By $A \Subset B$ we mean that the

© The Author(s), under exclusive license to Springer Nature Singapore Pte Ltd. 2023
R. Alicandro et al., *A Variational Theory of Convolution-Type Functionals*,
SpringerBriefs on PDEs and Data Science,
https://doi.org/10.1007/978-981-99-0685-7_2

closure of A is a compact subset of B. The Lebesgue d-dimensional measure and the Hausdorff $(d-1)$-dimensional measure in $\mathbb{R}^d$ are denoted by $\mathcal{L}^d$ and $\mathcal{H}^{d-1}$ respectively. We also denote the Lebesgue measure of a set $E \subset \mathbb{R}^d$ by $|E|$.

Given $g : \mathbb{R}^{m \times d} \to \mathbb{R}$ we use the notation

$$\nabla g(M) = \left(\frac{\partial g(M)}{\partial M_{i,j}}\right)_{\substack{i \in \{1,\ldots,m\} \\ j \in \{1,\ldots,d\}}} \in \mathbb{R}^{m \times d}$$

for the gradient of g. Given a matrix-valued map $\Sigma : \Omega \to \mathbb{R}^{m \times d}$ we let the *divergence* of Σ be the map $\mathrm{Div}(\Sigma) : \Omega \to \mathbb{R}^m$ with

$$\big(\mathrm{Div}(\Sigma)\big)_i = \mathrm{div}(\Sigma_i) \quad \text{for every } i = 1, \ldots, m,$$

where Σ_i denotes the i-th row of Σ.

We use standard notation for Lebesgue and Sobolev spaces, and their local versions. If u is an integrable function on a measurable set $E \subset \Omega$,

$$u_E := \frac{1}{|E|} \int_E u(x)dx$$

denotes the average of u on E. If $A \Subset B$, a *cut-off function* φ between A and B is a C^∞-function with $0 \leq \varphi \leq 1$, $\varphi = 0$ on ∂B and $\varphi = 1$ on A.

We use standard notation for Γ-convergence [1], indicating the topology with respect to which it is performed.

Unless otherwise stated, the letter C denotes a generic strictly positive constant independent of the parameters of the problem taken into account.

2.2 Setting of the Problem and Comments

Given $\varepsilon > 0$ and $f_\varepsilon : \Omega \times \mathbb{R}^d \times \mathbb{R}^m \to [0, +\infty)$ a positive Borel function, we introduce the non-local functional $F_\varepsilon : L^p(\Omega; \mathbb{R}^m) \to [0, +\infty]$ defined as

$$F_\varepsilon(u) := \int_{\mathbb{R}^d} \int_{\Omega_\varepsilon(\xi)} f_\varepsilon\left(x, \xi, \frac{u(x + \varepsilon\xi) - u(x)}{\varepsilon}\right) dx \, d\xi \tag{2.1}$$

where $\Omega_\varepsilon(\xi) := \{x \in \Omega : x + \varepsilon\xi \in \Omega\}$ (see Fig. 2.1).

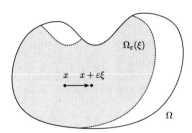

Fig. 2.1 The set $\Omega_\varepsilon(\xi)$

Note that definition (2.1) is equivalent to (1.15) up to the change of variable $y = x + \varepsilon\xi$. The reason to rewrite our energies in this apparently more complicated way is in order to state more clearly the hypotheses in the following, highlighting the 'microscopic interaction length' ξ. A particular class of functionals of the form (2.1), which will be used in the following as comparison energies, is introduced below.

Definition 2.1 (The *Convolution Functionals* $G_\varepsilon[a]$ and G_ε^r) Given a measurable function $a : \mathbb{R}^d \to [0, +\infty)$, we set

$$G_\varepsilon[a](u) := \int_{\mathbb{R}^d} a(\xi) \int_{\Omega_\varepsilon(\xi)} \left| \frac{u(x + \varepsilon\xi) - u(x)}{\varepsilon} \right|^p dx \, d\xi. \tag{2.2}$$

Moreover, we define the local versions $G_\varepsilon[a](u, A)$ as for F_ε in (2.3).

In the case in which $a = \chi_{B_r}$ we will simply write G_ε^r instead of $G_\varepsilon[\chi_{B_r}]$.

Remark 2.1 (Origin-Symmetric Kernels) The functionals $G_\varepsilon[a]$ do not change if we replace the kernel a with its symmetric part; that is,

$$G_\varepsilon[a](u) = G_\varepsilon[a^{sym}](u), \quad \text{where} \quad a^{sym}(\xi) := \frac{a(\xi) + a(-\xi)}{2}.$$

This can be easily seen by rewriting $G_\varepsilon[a]$ as in (1.6) and switching the roles of x and y.

We will study the behaviour of energies F_ε as $\varepsilon \to 0$, under proper assumptions on f_ε, by proving a compactness result with respect to Γ-convergence in the L^p-topology. In this context the role of convolution functionals $G_\varepsilon[a]$ will be the analog of that of the integral of the p-th power of the gradient as a comparison energy for local integral functional with p-growth.

For any $u \in L^p(\Omega; \mathbb{R}^m)$ and $A \in \mathcal{A}(\Omega)$ we also introduce a local version of the functionals in (2.1) by setting

$$F_\varepsilon(u, A) := \int_{\mathbb{R}^d} \int_{A_\varepsilon(\xi)} f_\varepsilon\left(x, \xi, \frac{u(x + \varepsilon\xi) - u(x)}{\varepsilon}\right) dx \, d\xi, \tag{2.3}$$

where $A_\varepsilon(\xi) := \{x \in A : x + \varepsilon\xi \in A\}$.

In what follows we use the standard notation

$$F'(u, A) := \Gamma\text{-}\liminf_{\varepsilon \to 0} F_\varepsilon(u, A), \quad F''(u, A) := \Gamma\text{-}\limsup_{\varepsilon \to 0} F_\varepsilon(u, A)$$

for the upper and lower Γ-limits (cf. [1]) performed with respect to the strong topology of $L^p(\Omega; \mathbb{R}^m)$.

Remark 2.2 Note that in this setting the upper and lower Γ-limits of $F_\varepsilon(\cdot, A)$ performed with respect to the strong $L^p(A; \mathbb{R}^m)$ topology or with respect to the strong $L^p(\Omega; \mathbb{R}^m)$ topology are the same. Indeed, for any sequence u_ε converging to $u \in L^p(\Omega; \mathbb{R}^m)$ strongly in $L^p(A; \mathbb{R}^m)$, we can take

$$\tilde{u}_\varepsilon(x) = \begin{cases} u_\varepsilon(x) & \text{if } x \in A \\ u(x) & \text{if } x \in \Omega \setminus A, \end{cases}$$

which converges to u strongly in $L^p(\Omega; \mathbb{R}^m)$ and leaves the energy unchanged; that is, $F_\varepsilon(u_\varepsilon, A) = F_\varepsilon(\tilde{u}_\varepsilon, A)$.

2.3 Assumptions

The main assumption on our energy densities f_ε introduced above is that they are controlled by convolution energies of p growth. In order to maintain the greatest generality and applicability of our results, we consider the following set of hypotheses: there exist two strictly positive constants r_0, c_0 and four families of non-negative functions $\psi_{\varepsilon,1}, \rho_{\varepsilon,1} : B_{r_0} \to [0, +\infty)$ and $\psi_{\varepsilon,2}, \rho_{\varepsilon,2} : \mathbb{R}^d \to [0, +\infty)$ satisfying

$$\limsup_{\varepsilon \to 0} \left(\int_{B_{r_0}} \rho_{\varepsilon,1}(\xi) \, d\xi + \int_{\mathbb{R}^d} \rho_{\varepsilon,2}(\xi) \, d\xi \right) < +\infty, \tag{2.4}$$

$$\limsup_{\varepsilon \to 0} \int_{\mathbb{R}^d} \psi_{\varepsilon,2}(\xi) |\xi|^p \, d\xi < +\infty, \tag{2.5}$$

$$\psi_{\varepsilon,1}(\xi) \geq c_0, \quad \text{for a.e. } \xi \in B_{r_0}, \tag{2.6}$$

if $\displaystyle\limsup_{\varepsilon \to 0} \int_{B_{r_0}} \psi_{\varepsilon,2}(\xi) \, d\xi = +\infty$ then there exists $c_1 > 0$ such that

$$\psi_{\varepsilon,2}(\xi) \leq c_1 \psi_{\varepsilon,1}(\xi), \text{ for a.e. } \xi \in B_{r_0} \tag{2.7}$$

such that

$$\psi_{\varepsilon,1}(\xi)|z|^p - \rho_{\varepsilon,1}(\xi) \leq f_\varepsilon(x, \xi, z) \quad \text{for a.e. } \xi \in B_{r_0}; \tag{H0}$$

$$f_\varepsilon(x, \xi, z) \leq \psi_{\varepsilon,2}(\xi)|z|^p + \rho_{\varepsilon,2}(\xi) \quad \text{for a.e. } \xi \in \mathbb{R}^d \tag{H1}$$

for a.e. $x \in \Omega$ and every $z \in \mathbb{R}^m$. Moreover for any $\delta > 0$ there exists $r_\delta > 0$ such that

$$\limsup_{\varepsilon \to 0} \int_{\mathbb{R}^d \setminus B_{r_\delta}} (\psi_{\varepsilon,2}(\xi)|\xi|^p + \rho_{\varepsilon,2}(\xi)) \, d\xi < \delta. \tag{H2}$$

Note that, when $f_\varepsilon(x, \xi, z) = a_\varepsilon(\xi)|z|^p$ the hypotheses above hold if a_ε satisfies (2.5), (2.6) and (H2). In the sequel, with a slight abuse of notation, we still denote with $\psi_{\varepsilon,1}$ and $\rho_{\varepsilon,1}$ their zero extension to the whole $\mathbb{R}^d$.

In what follows, when we refer to any of the assumptions (H0)–(H2) we will always implicitly assume that the functions $\rho_{\varepsilon,1}, \rho_{\varepsilon,2}, \psi_{\varepsilon,1}, \psi_{\varepsilon,2}$ satisfy (2.4)–(2.7).

For any $A \in \mathcal{A}(\Omega)$ and for all sufficiently small ε, we have that by hypotheses (H0) and (H1), the localized functionals satisfy

$$G_\varepsilon[\psi_{\varepsilon,1}](u, A) - C|A| \le F_\varepsilon(u, A) \le G_\varepsilon[\psi_{\varepsilon,2}](u, A) + C|A|. \tag{2.8}$$

By (2.6), the expression above implies

$$c_0 G_\varepsilon^{r_0}(u, A) - C|A| \le F_\varepsilon(u, A) \le G_\varepsilon[\psi_{\varepsilon,2}](u, A) + C|A|. \tag{2.9}$$

Remark 2.3 The hypotheses above can be compared with the standard p-growth assumptions for a family of integral functionals $\int_\Omega f_\varepsilon(x, \nabla u)\, dx$, which read

$$a_1(|z|^p - a_0) \le f_\varepsilon(x, z) \le a_2(1 + |z|^p) \tag{2.10}$$

for some positive constants a_0, a_1, and a_2. The polynomial lower-bound in (2.10) implies the boundedness of the L^p norm of the gradients of functions with equibounded energies and hence provides weak compactness in $W^{1,p}$ spaces. Analogously, we will show that condition (H0) and (2.6) provides strong compactness in L^p of functions with equibounded energies and yields that any limit function is in $W^{1,p}(\Omega; \mathbb{R}^m)$. Condition (H1) and (2.5) are the analog of the polynomial upper-bound in (2.10) and ensure that the Γ-limits are finite on $W^{1,p}$-functions. Condition (H2) is crucial to deduce the locality of the Γ-limits, in that it forbids relevant long-range interactions, and has no direct analog in terms of condition (2.10).

We spend now a few words to motivate also assumption (2.7) which, at a first glance, may appear mysterious. Loosely speaking, such a condition yields that the behaviour of the kernels $\psi_{\varepsilon,1}$ and $\psi_{\varepsilon,2}$ is the same for short-range interaction; i.e., as $|\xi| \to 0$. This ensures that any sequence with bounded energy u_ε, by (2.8) also satisfies $G_\varepsilon[\psi_{\varepsilon,2}](u_\varepsilon) \le c_1 G_\varepsilon[\psi_{\varepsilon,1}](u_\varepsilon) \le C$. Eventually, we notice that in the case in which the integral on a ball centered in zero of $\psi_{\varepsilon,2}$ is equibounded, the pointwise control $\psi_{\varepsilon,2} \le c_1 \psi_{\varepsilon,1}$ is not needed to bound $G_\varepsilon[\psi_{\varepsilon,2}]$ thanks to Lemma 4.1.

Remark 2.4 (Convolution Functionals with Kernels of Polynomial Decay) A simple class of kernels $\psi_{\varepsilon,1}, \psi_{\varepsilon,2}$ complying with (2.5)–(2.7) and (H2) are those satisfying

$$\psi_{\varepsilon,1}(\xi) = |\xi|^{-\alpha_1} \chi_{B_1}(\xi), \quad \psi_{\varepsilon,2}(\xi) = C \begin{cases} |\xi|^{-\alpha_2} & |\xi| < 1 \\ |\xi|^{-\beta_2} & |\xi| \ge 1 \end{cases}$$

for some $C > 1, 0 \leq \alpha_1 \leq \alpha_2 < p + d < \beta_2$ with

$$\alpha_1 = \begin{cases} 0 & \text{if } \alpha_2 < d \\ \alpha_2 & \text{otherwise.} \end{cases}$$

For the validity of the integral-representation result in Theorem 5.2 condition (H0) can be weakened requiring only that:

(H0′)   the computation of $F'(u, A)$ and $F''(u, A)$ can be restricted to families $\{u_\varepsilon\} \subset L^p(\Omega; \mathbb{R}^m)$ satisfying, for every open $A' \Subset A$,

$$c_0 G_\varepsilon^{r_0}(u_\varepsilon, A') \leq G_\varepsilon[\psi_{\varepsilon,1}](u_\varepsilon, A') \leq C(F_\varepsilon(u_\varepsilon, A) + |A|) \tag{2.11}$$

when ε is small enough, where C is a constant depending on $A \in \mathcal{A}(\Omega)$.

Such condition is clearly implied by (H0) and (2.6). Assuming (H0′) in place of (H0) allows to include in our analysis a wide varieties of cases which are not covered by (H0), such as convolution energies on perforated domains, where (2.11) is obtained by an extension theorem (see Sect. 6.4). In the two following examples we exhibit two further classes of convolution functionals satisfying (2.11).

Remark 2.5 (Control from Below Provided by a Translated Kernel) We now show that for the validity of (2.11) it suffices that $f_\varepsilon(x, \cdot, z)$ be controlled from below by an interaction kernel not necessarily centered in the origin. More precisely, assume that there exist $r_0, c_0 > 0$ and $\xi_0 \in \mathbb{R}^d$ such that

$$c_0(|z|^p - \rho_\varepsilon(\xi - \xi_0)) \leq f_\varepsilon(x, \xi, z) \quad \text{if } |\xi - \xi_0| \leq r_0 \tag{2.12}$$

(translated kernel). Take $r < r_0/2$ and an open $A' \Subset A$. For every $\eta \in B_r(\xi_0)$, by Jensen's inequality we have

$$G_\varepsilon^r(u, A') \leq 2^{p-1}\left(\int_{B_r} \int_{A'} \left| \frac{u(x + \varepsilon(\xi + \eta)) - u(x)}{\varepsilon} \right|^p dx\, d\xi \right.$$
$$\left. + \int_{B_r} \int_{A'} \left| \frac{u(x + \varepsilon(\xi + \eta)) - u(x + \varepsilon\xi)}{\varepsilon} \right|^p dx\, d\xi \right).$$

Note that, for ε small enough, the previous expression is well defined since $A' + \varepsilon(\xi + \eta) \subset A$, for every ξ and η as above. Averaging with respect to η we get

$$G_\varepsilon^r(u, A') \leq \frac{2^{p-1}}{|B_r|}\left(\int_{B_r(\xi_0)} \int_{B_r} \int_{A'} \left| \frac{u(x + \varepsilon(\xi + \eta)) - u(x)}{\varepsilon} \right|^p dx\, d\xi\, d\eta \right.$$
$$\left. + \int_{B_r(\xi_0)} \int_{B_r} \int_{A'} \left| \frac{u(x + \varepsilon(\xi + \eta)) - u(x + \varepsilon\xi)}{\varepsilon} \right|^p dx\, d\xi\, d\eta \right).$$

By the change of variable $\xi' = \xi + \eta$ and from the fact that $B_r(\eta) \subset B_{r_0}(\xi_0)$ we have

$$\frac{1}{|B_r|} \int_{B_r(\xi_0)} \int_{B_r} \int_{A'} \left| \frac{u(x + \varepsilon(\xi + \eta)) - u(x)}{\varepsilon} \right|^p dx\, d\xi\, d\eta$$

$$\leq \int_{B_{r_0}(\xi_0)} \int_{A'} \left| \frac{u(x + \varepsilon\xi') - u(x)}{\varepsilon} \right|^p dx\, d\xi' .$$

Using the change of variable $x' = x + \varepsilon\xi$ and the fact that $A' + \varepsilon\xi \subset A_\varepsilon(\eta)$ for every $\eta \in B_r(\xi_0)$, $\xi \in B_r$ we also get

$$\frac{1}{|B_r|} \int_{B_r(\xi_0)} \int_{B_r} \int_{A'} \left| \frac{u(x + \varepsilon(\xi + \eta)) - u(x + \varepsilon\xi)}{\varepsilon} \right|^p dx\, d\xi\, d\eta$$

$$\leq \int_{B_r(\xi_0)} \int_{A_\varepsilon(\eta)} \left| \frac{u(x' + \varepsilon\eta) - u(x')}{\varepsilon} \right|^p dx'\, d\eta .$$

Gathering all the inequalities above, for any open $A' \Subset A$ we obtain that

$$G_\varepsilon^r(u, A') \leq 2^p G_\varepsilon[\chi_{B_{r_0}(\xi_0)}](u, A)$$

for every ε small enough and (2.12) yields (2.11).

Remark 2.6 (Sign-Changing Kernels) In case of convolution(-type) energies the assumption that the kernels be non-negative can be relaxed to some extent. Note that this has no direct counterpart in condition (2.10) for integral functionals, since the negativeness of f_ε on a set of x of positive measure would give a Γ-limit identically equal to $-\infty$.

A simple example is obtained by taking r and r_0 as in the previous remark, $\xi_0 \in \mathbb{R}^d$ such that $|\xi_0| > 2r_0$ and $f_\varepsilon(x, \xi, z) = \left(\chi_{B_{r_0}(\xi_0)}(\xi) - \gamma \chi_{B_r(0)}(\xi) \right)|z|^p$, with $\gamma < 2^{-p}$. This function is negative for $|\xi| < r$. Nevertheless, by the previous remark, we obtain

$$(2^{-p} - \gamma) \int_{\mathbb{R}^d} \int_{B_r} \left| \frac{u(x + \varepsilon\xi) - u(x)}{\varepsilon} \right|^p d\xi\, dx$$

$$\leq \int_{\mathbb{R}^d} \int_{\mathbb{R}^d} f_\varepsilon\left(x, \xi, \frac{u(x + \varepsilon\xi) - u(x)}{\varepsilon}\right) d\xi\, dx,$$

which gives a bound for the convolution energies on the left-hand side for families with equibounded energies on the whole $\mathbb{R}^d$. From this bound it is possible to derive compactness properties. Note however that it does not immediately imply condition (2.11) for the localized energies, so that arguments requiring local estimates, such as integral-representation theorems, must be reworked.

We may also treat the case when the domain is not the whole $\mathbb{R}^d$ if we make some assumptions on the integration domain; e.g. convexity.

Let A be an open convex set. Applying Jensen's inequality and switching the roles of x and y we get

$$\int_{\{(x,y)\in A\times A\,:\,\varepsilon<|x-y|<2\varepsilon\}} \left|\frac{u(y)-u(x)}{\varepsilon}\right|^p dx\,dy$$

$$\leq 2^p \int_{\{(x,y)\in A\times A\,:\,\varepsilon<|x-y|<2\varepsilon\}} \left|\frac{u\left(\frac{x+y}{2}\right)-u(x)}{\varepsilon}\right|^p dx\,dy.$$

From the convexity of A, $(x+y)/2 \in A$ and $|(x+y)/2 - x| < \varepsilon$. Hence, by the change of variable $y' = (x+y)/2$ we obtain

$$\int_{\{(x,y)\in A\times A\,:\,\varepsilon<|x-y|<2\varepsilon\}} \left|\frac{u(y)-u(x)}{\varepsilon}\right|^p dx\,dy$$

$$\leq 2^{p+d} \int_{\{(x,y)\in A\times A\,:\,|x-y'|<\varepsilon\}} \left|\frac{u(y')-u(x)}{\varepsilon}\right|^p dx\,dy'.$$

Now if we consider

$$f_\varepsilon(x,\xi,z) = \left(2\chi_{B_1}(\xi) - \gamma\,\chi_{B_2\setminus B_1}(\xi)\right)|z|^p,$$

with $\gamma < 2^{-p-d}$, the previous computations yield $F_\varepsilon(u,A) \geq G_\varepsilon^1(u,A)$. This argument can be extended to sign-changing a such as the one pictured in Fig. 2.2 (cf. a_2); i.e., with $a(\xi)$ positive for small values of $|\xi|$. The arguments above are no longer true in the non-convex case and their generalization to every open set $A \in \mathcal{A}(\Omega)$ is not immediate.

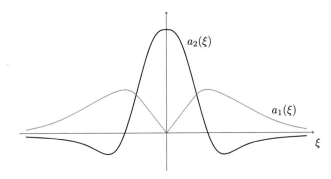

Fig. 2.2 Example of interaction kernels a_1 not complying with (2.6) and a_2 with changing sign

Remark 2.7 (A Technical Remark on Localization Techniques) Alternatively to (2.3), given an open set Ω, for any $u \in L^p(\Omega; \mathbb{R}^m)$ and $A \in \mathcal{A}(\Omega)$ we may introduce another (partially-)local version of the functionals in (2.1) by setting

$$\widetilde{F}_\varepsilon(u, A) := \int_{\mathbb{R}^d} \int_{A_\varepsilon(\Omega;\xi)} f_\varepsilon\left(x, \xi, \frac{u(x + \varepsilon\xi) - u(x)}{\varepsilon}\right) dx \, d\xi, \qquad (2.13)$$

where $A_\varepsilon(\Omega; \xi) := \{x \in A : x + \varepsilon\xi \in \Omega\}$. This version has the advantage of being additive in the set A, but loses the locality property (i.e., $\widetilde{F}_\varepsilon(u, A)$ depends also on the values of u on $\Omega \setminus A$), so that its Γ-limits must be computed with respect to the convergence in $L^p(\Omega; \mathbb{R}^m)$.

Reference

1. Braides, A.: Γ-Convergence for Beginners. Oxford Lecture Series in Mathematics and Its Applications, vol. 22. Oxford University Press, Oxford (2002)

Chapter 3
The Γ-Limit of a Class of Reference Energies

Abstract In this chapter we prove the Γ-convergence of convolution energies under suitable assumptions on the convolution kernels a_ε. In the scalar case, a general Γ-convergence result for functionals of this type has been proven in Ponce (Calc. Var. Partial Differ. Equ. 19:229–255, 2004). We give a new proof of the Γ-convergence which is independent of the arguments in that work. Such a result on the one hand shows a non-trivial example of Γ-converging functionals, on the other hand it allows, by comparison, to deduce that the Γ-limits of a family of functionals F_ε satisfy standard p-growth assumptions in a Sobolev-space setting.

Keywords Γ-convergence · Convolution energies · Radially symmetric kernels · Slicing method

Since our families of non-local functionals F_ε can be estimated by convolution energies, we first examine the asymptotic behaviour of the latter. For the sake of brevity, in this chapter we use the notation $\hat{\xi} = \frac{\xi}{|\xi|}$.

3.1 The Γ-Limit of $G_\varepsilon[a_\varepsilon]$

We focus on a family of functionals $G_\varepsilon[a_\varepsilon]$, in the notation introduced in Definition 2.1. Throughout this chapter if $a_\varepsilon : \mathbb{R}^d \to [0, +\infty)$ is a measurable function with finite p-moments we use the notation μ_a for the measure

$$\mu_a(E) = \int_E a(\xi)|\xi|^p \, d\xi \tag{3.1}$$

where $E \subseteq \mathbb{R}^d$ is a measurable set. The following result generalizes a theorem by Ponce [3].

Theorem 3.1 *Let $A \in \mathcal{A}^{\mathrm{reg}}(\Omega)$ and let $a_\varepsilon : \mathbb{R}^d \to [0, +\infty)$ be a family of non-negative measurable functions such that*

$$\limsup_{\varepsilon \to 0} \int_{\mathbb{R}^d} a_\varepsilon(\xi) |\xi|^p \, d\xi < +\infty \qquad (3.2)$$

and such that for every $\delta > 0$ there exists r_δ satisfying

$$\limsup_{\varepsilon \to 0} \int_{\mathbb{R}^d \setminus B_{r_\delta}} a_\varepsilon(\xi) |\xi|^p d\xi < \delta. \qquad (3.3)$$

Suppose that the measures μ_{a_ε} weakly* converge to a measure μ such that $\mathrm{span}(\mathrm{supp}(\mu)) = \mathbb{R}^d$. Then

$$\Gamma(L^p)\text{-}\lim_{\varepsilon \to 0} G_\varepsilon[a_\varepsilon](u, A) = \begin{cases} \int_A \int_{\mathbb{R}^d} |\nabla u(x)\hat{\xi}|^p \, d\mu(\xi) \, dx & \text{if } u \in W^{1,p}(A; \mathbb{R}^m) \\ +\infty & \text{otherwise.} \end{cases} \qquad (3.4)$$

Proof It is a straightforward consequence of Propositions 3.1 and 3.2 below.   □

We draw the attention of the reader on the fact that assumption (3.2) implies the weak*-compactness of the sequence $\{\mu_{a_\varepsilon}\}$ as $\varepsilon \to 0$ and therefore the additional hypothesis that the measures μ_{a_ε} weakly* converge to μ can be removed up to passing to subsequences.

Example 3.1 (Radially-Symmetric Kernels: Scalar Case) If $d\mu(\xi) = a(|\xi|)|\xi|^p \, d\xi$ and $m = 1$ then the limit energy is simply $c_a \int_A |\nabla u(x)|^p \, dx$, with

$$c_a = \int_{\mathbb{R}^d} a(\xi) |\xi_1|^p \, d\xi. \qquad (3.5)$$

Indeed, let $v_x \in \mathbb{S}^{d-1}$ be such that $\nabla u(x) = |\nabla u(x)| v_x$ for a.e. $x \in A$ and let $R_x \in SO(d)$ the rotation such that $R_x v_x = e_1$. Then we have

$$\int_{\mathbb{R}^d} a(|\xi|)|\xi|^p \int_A |\langle \nabla u(x), \hat{\xi} \rangle|^p \, dx \, d\xi = \int_{\mathbb{R}^d} \int_A a(|\xi|)|\langle e_1, R_x \xi \rangle|^p |\nabla u(x)|^p \, dx \, d\xi$$

and the change of variable $\xi' = R_x \xi$ yields the claim. In particular, if $a = \chi_{B_R}$ then $c_a = c_R = \int_{B_R} |\xi_1|^p \, d\xi$.

In the general vectorial case $m \geq 1$ with $p = 2$ we easily infer the same conclusion (see also Example 6.1).

Remark 3.1 We can rewrite the right-hand side of (3.4) for $u \in W^{1,p}(A; \mathbb{R}^m)$ as

$$\int_A f_0(\nabla u(x)) \, dx,$$

where $f_0(M) := \int_{\mathbb{R}^d} |M\hat{\xi}|^p \, d\mu(\xi)$. The function $f_0 : \mathbb{R}^{m \times d} \to [0, +\infty)$ is clearly convex. Note that it satisfies

$$C_0|M|^p \leq f_0(M) \leq C_1|M|^p \tag{3.6}$$

for some $C_1 \geq C_0 > 0$ depending on μ.

Indeed, (3.2), the lower-semicontinuity of the total variation with respect to weak*-convergence and the fact that supp(μ) is nontrivial imply that $0 < \mu(\mathbb{R}^d) < +\infty$, yielding the right-hand side of (3.6). For the lower bound, let $\xi_1, \ldots, \xi_d \in$ supp(μ) be independent vectors. Then, for every $r > 0$ and every $\xi \in B_r(\xi_j)$ we get

$$|M\xi_j|^p \leq 2^{p-1}|M\xi|^p + 2^{p-1}|M|^p r^p,$$

for every $j = 1, \ldots, d$. Thus, for sufficiently small r we obtain

$$f_0(M) \geq C \sum_{j=1}^{d} \mu(B_r(\xi_j)) \left(\frac{1}{2^{p-1}} |M\hat{\xi}_j|^p - |M|^p r^p \right) \geq C_0|M|^p,$$

where the last inequality comes from the fact that $\|M\|_\mu := \left(\sum_{j=1}^{d} |M\hat{\xi}_j|^p \right)^{\frac{1}{p}}$ is a norm on $\mathbb{R}^{m \times d}$.

Remark 3.2 Rereading Theorem 3.1 in terms of the notation of Sect. 2.3, assumption (3.2) corresponds to (2.5) and ensures that $\Gamma(L^p)\text{-}\lim_{\varepsilon \to 0} G_\varepsilon[a_\varepsilon](u, A)$ is finite on $W^{1,p}(A; \mathbb{R}^m)$. Assumption (2.6) is replaced by the more general condition span(supp(μ)) $= \mathbb{R}^d$. Moreover, observe that (3.3) corresponds to assumption (H2), which is crucial to ensure the locality of the Γ-limit, as shown by the example below.

Example 3.2 (Effects of the Failure of the Locality Assumption) Given $\Omega = (0, 1)$, consider the functions

$$a_\varepsilon(\xi) = \begin{cases} 1 & |\xi| \leq 1 \\ \varepsilon^p & \frac{1}{2\varepsilon} - 1 < \xi < \frac{1}{2\varepsilon} + 1 \\ 0 & \text{otherwise.} \end{cases}$$

Such kernels do not satisfy (3.3). Correspondingly, the densities $f_\varepsilon(x, \xi, z) = a_\varepsilon(\xi)|z|^p$ satisfy hypotheses (H0), (H1) and conditions (2.5) but not the locality assumption (H2). Now, for any given $u \in W^{1,p}(0, 1)$, and any u_ε converging to u in $L^p(0, 1)$ as $\varepsilon \to 0$, we have

$$G_\varepsilon[a_\varepsilon](u_\varepsilon) = \int_{-\infty}^{+\infty} a_\varepsilon(\xi) \int_{0 \vee (-\varepsilon\xi)}^{1 \wedge (1-\varepsilon\xi)} \left| \frac{u_\varepsilon(x + \varepsilon\xi) - u_\varepsilon(x)}{\varepsilon} \right|^p dx \, d\xi$$

$$= G_\varepsilon^1(u_\varepsilon) + \int_{\frac{1}{2\varepsilon}-1}^{\frac{1}{2\varepsilon}+1} \int_0^{1-\varepsilon\xi} |u_\varepsilon(x + \varepsilon\xi) - u_\varepsilon(x)|^p dx \, d\xi.$$

Taking the limit as $\varepsilon \to 0$, by the stability of Γ-convergence with respect to continuous perturbations, Theorem 3.1 and Lebesgue's Dominated Convergence Theorem we get

$$\Gamma\text{-}\lim_{\varepsilon \to 0} G_\varepsilon[a_\varepsilon](u) = \frac{2}{p+1} \int_0^1 |u'(x)|^p dx + 2 \int_0^{\frac{1}{2}} \left| u\left(x + \frac{1}{2}\right) - u(x) \right|^p dx.$$

The limit functional is still defined on $W^{1,p}(0,1)$ but does not have a local representation.

In the following two lemmas we deal with the Γ-lim inf and the Γ-lim sup of the family of functionals $G_\varepsilon[a_\varepsilon]$ separately.

Proposition 3.1 *Let $a_\varepsilon : \mathbb{R}^d \to [0, +\infty)$ be a family of non-negative measurable functions with finite p-moments such that the measures μ_{a_ε}, defined in (3.1), weakly* converge to a measure μ such that* span(supp(μ)) $= \mathbb{R}^d$. *Then for every $u \in L^p(\Omega; \mathbb{R}^m)$ and $A \in \mathcal{A}(\Omega)$*

$$\Gamma\text{-}\liminf_{\varepsilon \to 0} G_\varepsilon[a_\varepsilon](u, A) \geq \begin{cases} \int_A \int_{\mathbb{R}^d} |\nabla u(x)\hat{\xi}|^p \, d\mu(\xi) \, dx & \text{if } u \in W^{1,p}(A; \mathbb{R}^m) \\ +\infty & \text{otherwise.} \end{cases}$$

$$(3.7)$$

Proof When needed, we will identify the functions with their extensions equal to 0 outside Ω. We will use a slicing procedure, so we first deal with the one-dimensional case; i.e., when $\Omega \subset \mathbb{R}$ and, for the sake of simplicity, $A = (0, 1)$. In this case the energy reads as

$$G_\varepsilon[a_\varepsilon](u, A) = \int_{-\infty}^{+\infty} a_\varepsilon(\xi) \int_{0 \vee (-\varepsilon\xi)}^{(1-\varepsilon\xi) \wedge 1} \left| \frac{u(x + \varepsilon\xi) - u(x)}{\varepsilon} \right|^p dx \, d\xi.$$

For every fixed $\xi > 0$ we set $N = \lfloor 1/(\varepsilon\xi) \rfloor - 1$ and applying Jensen's inequality we have

$$\int_0^{1-\varepsilon\xi} \left| \frac{u(x + \varepsilon\xi) - u(x)}{\varepsilon} \right|^p dx \geq \sum_{k=0}^{N-1} \int_{k\varepsilon\xi}^{(k+1)\varepsilon\xi} \left| \frac{u(x + \varepsilon\xi) - u(x)}{\varepsilon} \right|^p dx$$

$$\geq \sum_{k=0}^{N-1} \varepsilon\xi \left| \frac{1}{\varepsilon\xi} \int_{k\varepsilon\xi}^{(k+1)\varepsilon\xi} \frac{u(x + \varepsilon\xi) - u(x)}{\varepsilon} dx \right|^p.$$

$$(3.8)$$

For every $k \in \mathbb{Z}$ set

$$u_{\varepsilon\xi}^{(k)} = \frac{1}{\varepsilon\xi} \int_{k\varepsilon\xi}^{(k+1)\varepsilon\xi} u(x) dx$$

and let $u_{\varepsilon\xi} : \mathbb{R} \to \mathbb{R}^m$ be the piecewise-affine function that interpolates the values $\{u(\varepsilon\xi)^{(k)}\}_{k\in\mathbb{Z}}$ on the lattice $\varepsilon\xi\mathbb{Z}$. Thus inequality (3.8) reads

$$\int_0^{1-\varepsilon\xi} \left|\frac{u(x+\varepsilon\xi)-u(x)}{\varepsilon}\right|^p dx \geq \sum_{k=0}^{N-1} \varepsilon\xi \left|\frac{u_{\varepsilon\xi}^{(k+1)} - u_{\varepsilon\xi}^{(k)}}{\varepsilon}\right|^p$$

$$= \xi^p \int_0^{N\varepsilon\xi} |u'_{\varepsilon\xi}(x)|^p dx.$$

The same analysis when $\xi < 0$ implies

$$\int_{0\vee(-\varepsilon\xi)}^{(1-\varepsilon\xi)\wedge 1} \left|\frac{u(x+\varepsilon\xi)-u(x)}{\varepsilon}\right|^p dx \geq |\xi|^p \int_{2\varepsilon|\xi|}^{1-2\varepsilon|\xi|} |u'_{\varepsilon|\xi|}(x)|^p dx,$$

for every $\xi \in \mathbb{R}$. We can generalize the previous argument to any $A \subset \Omega$ open subset obtaining

$$\int_{A_\varepsilon(\xi)} \left|\frac{u(x+\varepsilon\xi)-u(x)}{\varepsilon}\right|^p dx \geq |\xi|^p \int_{\tilde{A}_\varepsilon(\xi)} |u'_{\varepsilon|\xi|}(x)|^p dx, \tag{3.9}$$

for every $\xi \in \mathbb{R}$, where $\tilde{A}_\varepsilon(\xi) = \{x \in A : \text{dist}(x, A^c) > 2\varepsilon|\xi|\}$.

We now extend (3.9) to any dimension by a slicing method. For any $\xi \in \mathbb{R}^d\setminus\{0\}$ define $\Pi_\xi = \{y \in \mathbb{R}^d : \langle y, \xi\rangle = 0\}$, and for any $y \in \Pi_\xi$ set

$$\Omega_{y,\xi} = \{s \in \mathbb{R} : y + s\hat{\xi} \in \Omega\}, \qquad A_{y,\xi} = \{s \in \mathbb{R} : y + s\hat{\xi} \in A\},$$

$$A_{y,\xi}^\varepsilon = \{s \in \mathbb{R} : y + s\hat{\xi} \in A_\varepsilon(\xi)\}, \quad \tilde{A}_{y,\xi}^\varepsilon = \{s \in \mathbb{R} : \text{dist}(s, A_{y,\xi}^c) > 2\varepsilon|\xi|\},$$

$$u_{y,\xi}(s) := u(y + s\hat{\xi}) \quad s \in \mathbb{R}. \tag{3.10}$$

Fubini's Theorem yields

$$\int_{A_\varepsilon(\xi)} \left|\frac{u(x+\varepsilon\xi)-u(x)}{\varepsilon}\right|^p dx = \int_{\Pi_\xi} \int_{A_{y,\xi}^\varepsilon} \left|\frac{u_{y,\xi}(s+\varepsilon|\xi|)-u_{y,\xi}(s)}{\varepsilon}\right|^p ds\, dy.$$

By (3.9) we get

$$\int_{A_\varepsilon(\xi)} \left|\frac{u(x+\varepsilon\xi)-u(x)}{\varepsilon}\right|^p dx \geq |\xi|^p \int_{\Pi_\xi} \int_{\tilde{A}_{y,\xi}^\varepsilon} |u'_{y,\varepsilon,\xi}(s)|^p ds\, dy, \tag{3.11}$$

where $u_{y,\varepsilon,\xi} : \mathbb{R} \to \mathbb{R}^m$ denotes the piecewise-affine function interpolating the values $\{(u_{y,\xi})_{\varepsilon|\xi|}^{(k)}\}_{k\in\mathbb{Z}}$ on $\varepsilon|\xi|\mathbb{Z}$.

Let $\{\varepsilon_j\}$ be a sequence converging to 0 and let $u_j \to u$ in $L^p(\Omega; \mathbb{R}^m)$. Without loss of generality we may assume that $G_{\varepsilon_j}[a_{\varepsilon_j}](u_j, A)$ is uniformly bounded. Let

$(u_j)_{y,\xi}$ denote the corresponding slicing functions defined by (3.10). Then we have that $(u_j)_{y,\xi} \to u_{y,\xi}$ in $L^p(\Omega_{y,\xi}; \mathbb{R}^m)$. Moreover, by Braides [2, Lemma 3.36] $(u_j)_{y,\varepsilon_j,\xi} \to u_{y,\xi}$ in $L^p(\Omega_{y,\xi}; \mathbb{R}^m)$ for almost every $y \in \Pi_\xi$. We set

$$\varphi_j(\xi) := \int_{\Pi_\xi} \int_{\tilde{A}^{\varepsilon_j}_{y,\xi}} |(u_j)'_{y,\varepsilon_j,\xi}(s)|^p ds \, dy,$$

and claim that for every $\xi \in E$ we have

$$\liminf_{j \to \infty} \varphi_j(\xi) \geq \int_{\Pi_\xi} \int_{A_{y,\xi}} |u'_{y,\xi}(s)|^p ds \, dy. \tag{3.12}$$

Indeed, this is trivial when the left-hand side is $+\infty$. If, otherwise, the left-hand side is finite then for almost every $y \in \Pi_\xi$ up to subsequences $(u_j)'_{y,\varepsilon_j,\xi}$ is bounded in L^p and so $(u_j)_{y,\varepsilon_j,\xi}$ weakly converges to $u_{y,\xi}$ in $W^{1,p}(A_{y,\xi}; \mathbb{R}^m)$. Then, Fatou's Lemma and the lower semicontinuity of the L^p-norm with respect to the weak convergence yields (3.12).

Now, we use Egorov's and Lusin's theorems to improve the convergence of φ_j to a uniform convergence so that it can be coupled with the weak* convergence of the kernels a_{ε_j}. For this we also need a truncation argument. Let $M > 0$ be fixed. We drop the dependence on M for simplicity of notation. Set

$$\tilde{\varphi}_j(\xi) = \min\Big\{ M, \inf_{k \geq j} \varphi_k(\xi) \Big\},$$

which converges almost everywhere to

$$\varphi(\xi) = \min\Big\{ M, \liminf_{j \to +\infty} \varphi_j(\xi) \Big\}.$$

By Egorov's theorem for every $\delta > 0$ there exists $E_\delta \subset \mathbb{R}^d$ with $\mu(\mathbb{R}^d \setminus E_\delta) < \delta$ such that $\tilde{\varphi}_j \chi_{E_\delta}$ converge uniformly to $\varphi \chi_{E_\delta}$ on $\mathbb{R}^d$. By Lusin's theorem (see for instance [1, Theorem 1.45 and Remark 1.46]) there exist $v_j^\delta \in C_c(\mathbb{R}^d; [0, M])$ and compacts K_j^δ with $\mu(\mathbb{R}^d \setminus K_j^\delta) < 2^{-j}\delta$ such that $v_j^\delta = \tilde{\varphi}_j \chi_{E_\delta}$ on K_j^δ. Hence, $v_j^\delta \equiv \tilde{\varphi}_j \chi_{E_\delta}$ for every j on the compact $K^\delta = \bigcap_{j=1}^\infty K_j^\delta$ and $\mu(\mathbb{R}^d \setminus K^\delta) < \delta$. This implies that v_j^δ converge uniformly on K^δ to $\varphi \chi_{E_\delta}$. By the weak* convergence of μ_{a_ε} and (3.12) we get

$$\lim_{j \to +\infty} \int_{\mathbb{R}^d} a_{\varepsilon_j}(\xi) |\xi|^p \tilde{\varphi}_j(\xi) \chi_{E_\delta}(\xi) \, d\xi$$

$$\geq \int_{K^\delta \cap E_\delta} \varphi(\xi) \, d\mu(\xi)$$

$$\geq \int_{K^\delta \cap E_\delta} \min\Big\{ M, \int_{\Pi_\xi} \int_{A_{y,\xi}} |u'_{y,\xi}(s)|^p ds \, dy \Big\} d\mu(\xi).$$

Multiplying both members of (3.11), with u_j in place of u, by $a_{\varepsilon_j}(\xi)$, integrating in ξ and then taking the limit as $j \to +\infty$, we obtain

$$\liminf_{j\to+\infty} G_{\varepsilon_j}[a_{\varepsilon_j}](u_j, A) \geq \int_{\mathbb{R}^d} \int_{\Pi_\xi} \int_{A_{y,\xi}} |u'_{y,\xi}(s)|^p \, ds \, dy \, d\mu(\xi) \qquad (3.13)$$

by the arbitrariness of δ and M. If $u \in W^{1,p}(A; \mathbb{R}^m)$, for almost every $\xi \in \mathbb{R}^d \setminus \{0\}$ and $y \in \Pi_\xi$

$$u'_{y,\xi}(s) = \nabla u(x)\hat{\xi}, \quad x = y + s\hat{\xi}$$

then Fubini's Theorem leads to (3.7). Thus it remains to prove that if the left-hand side of (3.13) is finite then $u \in W^{1,p}(A; \mathbb{R}^m)$. From (3.13)

$$\int_{\Pi_\xi} \int_{A_{y,\xi}} |u'_{y,\xi}(s)|^p \, ds \, dy < +\infty \qquad (3.14)$$

for μ-a.e. $\xi \in \mathbb{R}^d$. Since $\mathrm{span}(\mathrm{supp}(\mu)) = \mathbb{R}^d$, (3.14) holds for linearly independent points $\xi_1, \ldots, \xi_d \in \mathbb{R}^d$, from which the conclusion follows. □

Proposition 3.2 Let $a_\varepsilon : \mathbb{R}^d \to [0, +\infty)$ be a family of non-negative functions satisfying (3.2). Then, for every $A \in \mathcal{A}^{\mathrm{reg}}(\Omega)$ and $u \in W^{1,p}(A; \mathbb{R}^m)$

$$\Gamma\text{-}\limsup_{\varepsilon\to0} G_\varepsilon[a_\varepsilon](u, A) \leq C \int_A |\nabla u(x)|^p \, dx, \qquad (3.15)$$

for some constant $C > 0$. Suppose in addition that, for every $\delta > 0$ there exists r_δ such that (3.3) holds and that the measures μ_{a_ε} as in (3.1) weakly* converge to a measure μ such that $\mathrm{span}(\mathrm{supp}(\mu)) = \mathbb{R}^d$. Then, for every $A \in \mathcal{A}^{\mathrm{reg}}(\Omega)$ and $u \in W^{1,p}(A; \mathbb{R}^m)$

$$\Gamma\text{-}\limsup_{\varepsilon\to0} G_\varepsilon[a_\varepsilon](u, A) \leq \int_A \int_{\mathbb{R}^d} |\nabla u(x)\hat{\xi}|^p \, d\mu(\xi) \, dx. \qquad (3.16)$$

Proof By a density argument, we may restrict to the case $u \in C_c^\infty(\mathbb{R}^d; \mathbb{R}^m)$. By Remark 4.1 (see Chapter 4), in order to prove (3.15) it is sufficient to estimate the upper limit of $G_\varepsilon[a_\varepsilon \chi_{B_T}](u, A)$, for some $T > 0$. For every $x \in \mathbb{R}^d$ we have

$$\frac{u(x + \varepsilon\xi) - u(x)}{\varepsilon} = \int_0^1 \nabla u(x + s\varepsilon\xi)\xi \, ds,$$

and by Jensen's inequality and Fubini's Theorem we get

$$G_\varepsilon[a_\varepsilon \chi_{B_T}](u, A) \leq \int_{B_T} a_\varepsilon(\xi)|\xi|^p \int_{A_\varepsilon(\xi)} \int_0^1 |\nabla u(x + s\varepsilon\xi)|^p \, ds \, dx \, d\xi$$

$$\leq \int_{B_T} a_\varepsilon(\xi)|\xi|^p \int_0^1 \int_{A+B_{\varepsilon T}} |\nabla u(x)|^p dx\, ds\, d\xi$$

$$\leq \int_{B_T} a_\varepsilon(\xi)|\xi|^p \int_A |\nabla u(x)|^p dx\, d\xi + o(1).$$

Taking the lim sup as $\varepsilon \to 0$, by (3.2) we obtain (3.15).

We now prove (3.16) under the additional assumption (3.3) and the weak* convergence of μ_{a_ε} to μ. We split $G_\varepsilon[a_\varepsilon](u, A)$ as follows

$$G_\varepsilon[a_\varepsilon](u, A) = \int_{B_{r_\delta}} a_\varepsilon(\xi) \int_{A_\varepsilon(\xi)} \left| \frac{u(x+\varepsilon\xi) - u(x)}{\varepsilon} \right|^p dx\, d\xi$$

$$+ \int_{B_{r_\delta}^c} a_\varepsilon(\xi) \int_{A_\varepsilon(\xi)} \left| \frac{u(x+\varepsilon\xi) - u(x)}{\varepsilon} \right|^p dx\, d\xi$$

for any $\delta > 0$, where r_δ is defined as in assumption (3.3). Expanding $u(x)$ at the first order when $|\xi| < r_\delta$ we get

$$\int_{B_{r_\delta}} a_\varepsilon(\xi) \int_{A_\varepsilon(\xi)} \left| \frac{u(x+\varepsilon\xi) - u(x)}{\varepsilon} \right|^p dx\, d\xi$$

$$= \int_{B_{r_\delta}} a_\varepsilon(\xi) \left(\int_{A_\varepsilon(\xi)} |\nabla u(x)\xi|^p dx + o(1) \right) d\xi$$

and for $|\xi| > r_\delta$ by assumption (3.3)

$$\int_{B_{r_\delta}^c} a_\varepsilon(\xi) \int_{A_\varepsilon(\xi)} \left| \frac{u(x+\varepsilon\xi) - u(x)}{\varepsilon} \right|^p dx\, d\xi$$

$$\leq \int_{B_{r_\delta}^c} a_\varepsilon(\xi) \int_A (\|\nabla u\|_{L^\infty(\mathbb{R}^d)} |\xi|)^p dx\, d\xi \leq |A| \|\nabla u\|_{L^\infty(\mathbb{R}^d)}^p \delta.$$

Hence, gathering the inequalities above we obtain

$$G_\varepsilon[a_\varepsilon](u, A) \leq \int_{B_{r_\delta}} a_\varepsilon(\xi) \int_A |\nabla u(x)\xi|^p dx\, d\xi + C\delta \tag{3.17}$$

for some $C > 0$. Letting $\varepsilon \to 0$ in (3.17), we get

$$\limsup_{\varepsilon \to 0} G_\varepsilon[a_\varepsilon](u, A) \leq \int_A \int_{B_{r_\delta}} |\nabla u(x)\hat\xi|^p d\mu(\xi)\, dx + C\delta$$

and the arbitrariness of δ implies (3.16).   □

From the right-hand side inequality in (2.9), (2.11), Propositions 3.1 and 3.2 the following estimates hold.

Proposition 3.3 *Given $A \in \mathcal{A}^{\mathrm{reg}}(\Omega)$, let $\{F_\varepsilon(\cdot, A)\}$ be the family of functionals defined by (2.3) and assume that (H0'), (H1) and (H2) hold. If $F'(u, A)$ is finite, then $u \in W^{1,p}(A; \mathbb{R}^m)$. Moreover for every $u \in W^{1,p}(A; \mathbb{R}^m)$ we have*

$$F'(u, A) \geq c\,(\|\nabla u\|^p_{L^p(A)} - |A|), \qquad (3.18)$$

$$F''(u, A) \leq C(\|\nabla u\|^p_{L^p(A)} + |A|), \qquad (3.19)$$

for some positive constants c, C.

References

1. Ambrosio, L., Fusco, N., Pallara, D.: Functions of Bounded Variation and Free Discontinuity Problems. Oxford Mathematical Monographs. Oxford University Press, New York (2000)
2. Braides, A.: Approximation of Free-Discontinuity Problems. Lecture Notes in Mathematics, vol. 1694. Springer-Verlag, Berlin (1998)
3. Ponce, A.C.: A new approach to Sobolev spaces and connections to Γ-convergence. Calc. Var. Partial Differ. Equ. **19**, 229–255 (2004)

Chapter 4
Asymptotic Embedding and Compactness Results

Abstract In this section we include some results that extend corresponding results in Sobolev spaces to the case of convolution energies. In Theorem 4.2 we prove the compactness of sequences of functions for which both the L^p norms and the energies are uniformly bounded, whose proof is a non-local counterpart of that of the classical Riesz-Fréchet-Kolmogorov Theorem. We will see that the role played by the L^p norm of gradients in the standard compact immersion results of Sobolev spaces is played in our context by the energies G_ε^r. In Propositions 4.1 and 4.2 we show the validity of Poincaré-type inequalities.

Keywords Extension theorem · Short-range interactions · Long-range interactions · L^p-compactness · Poincaré-type inequalities

Before proving compactness results in the next sections, we provide some preliminary results of independent interest concerning extension operators and the possibility of controlling long-range interactions with short-range interactions. Subsequently, we prove some asymptotic analogs of embedding and compactness results for Sobolev spaces (see e.g. [2]).

4.1 An Extension Result

By mimicking a standard procedure of extension of Sobolev functions, we provide the following corresponding result.

Theorem 4.1 (Extension) *Let A be any open set with Lipschitz boundary with ∂A bounded and let r > 0. Then there exist an open set $\tilde{A} \supseteq A$, a linear continuous map*

$$E : L^p(A; \mathbb{R}^m) \to L^p(\tilde{A}; \mathbb{R}^m)$$

and three positive constants $C = C(A), r_1 = r_1(r, A), \varepsilon_0 = \varepsilon_0(r, A)$ such that for all $u \in L^p(A; \mathbb{R}^m)$,

$$Eu = u \quad \text{a.e. in} \quad A,$$

$$\|Eu\|^p_{L^p(\tilde{A};\mathbb{R}^m)} + G^{r_1}_\varepsilon(Eu, \tilde{A}) \leq C\left(\|u\|^p_{L^p(A;\mathbb{R}^m)} + G^r_\varepsilon(u, A)\right)$$

for all $\varepsilon < \varepsilon_0$.

Proof We follow the construction of the extension of functions in fractional Sobolev spaces (see [1]). From the boundedness and the Lipschitz regularity of the boundary we can find a finite open covering $\{U_i\}_{i=1}^n$ of ∂A and Lipschitz invertible maps $H_i : Q \to U_i$ with $\|DH_i\|_{L^\infty(Q)} \leq L$ and $\|DH_i^{-1}\|_{L^\infty(U_i)} \leq L$, such that

$$H_i(Q^+) = A \cap U_i, \quad H_i(Q^-) = A^c \cap U_i, \quad H_i(Q_0) = \partial A \cap U_i,$$

where $Q = (-1, 1)^d$, $Q^\pm = \{x \in Q : \pm x_1 > 0\}$ and $Q^0 = \{x \in Q : x_1 = 0\}$. Let $\{\varphi_i\}_{i=0}^n \subset C_0^\infty(\mathbb{R}^d)$ be a partition of the unity, that is $V_i := \text{supp}(\varphi_i) \Subset U_i$ and

$$\sum_{i=0}^n \varphi_i(x) = 1, \text{ for every } x \in A,$$

where $A \backslash \bigcup_{i=1}^n U_i \subset U_0 \subset A$ is an open set. Define the maps $R_i : A^c \cap U_i \to A \cap U_i$ as follows

$$R_i(x) = H_i(-y_1, y_2, \ldots, y_d),$$

where $H_i^{-1}(x) = y = (y_1, \ldots, y_d) \in Q^-$. Note that R_i are invertible Lipschitz maps of Lipschitz constant less than L^2. Given $u \in L^p(A; \mathbb{R}^m)$, define

$$u_i(x) := \begin{cases} u(x) & x \in A \cap U_i \\ u(R_i(x)) & x \in A^c \cap U_i, \end{cases} \qquad \tilde{u}_i(x) := \varphi_i(x)u_i(x).$$

Then $\tilde{u} := \sum_{i=0}^n \tilde{u}_i$ extends u on $\tilde{A} = \bigcup_{i=0}^n U_i$. For every $0 \leq i \leq n$ we have

$$\int_{U_i} |u_i(x)|^p dx \leq C \int_{U_i \cap A} |u(x)|^p dx, \tag{4.1}$$

where $C > 0$ depends only on A, which yields $\|\tilde{u}\|_{L^p(\tilde{A};\mathbb{R}^m)}^p \leq C\|u\|_{L^p(A;\mathbb{R}^m)}^p$. Now, we show that for every $1 \leq i \leq n$

$$G_\varepsilon^{r_1}(u_i, U_i) \leq C G_\varepsilon^r(u, U_i \cap A) \qquad (4.2)$$

with $r_1 = r/(1 + 2L^2)$, which is trivial when $i = 0$. Using the change of variable $y = x + \varepsilon\xi$, it is convenient to rewrite the energy functionals as

$$G_\varepsilon^{r_1}(u_i, U_i) = \frac{1}{\varepsilon^d} \int_{U_i} \int_{U_i \cap B_{\varepsilon r_1}(x)} \left| \frac{u_i(y) - u_i(x)}{\varepsilon} \right|^p dy \, dx$$

$$\leq G_\varepsilon^r(u, U_i \cap A) \qquad (4.3)$$

$$+ \frac{1}{\varepsilon^d} \int_{U_i \cap A} \int_{(U_i \backslash A) \cap B_{\varepsilon r_1}(x)} \left| \frac{u_i(y) - u_i(x)}{\varepsilon} \right|^p dy \, dx \quad (4.4)$$

$$+ \frac{1}{\varepsilon^d} \int_{U_i \backslash A} \int_{U_i \cap A \cap B_{\varepsilon r_1}(x)} \left| \frac{u_i(y) - u_i(x)}{\varepsilon} \right|^p dy \, dx \quad (4.5)$$

$$+ \frac{1}{\varepsilon^d} \int_{U_i \backslash A} \int_{(U_i \backslash A) \cap B_{\varepsilon r_1}(x)} \left| \frac{u_i(y) - u_i(x)}{\varepsilon} \right|^p dy \, dx. \quad (4.6)$$

For every $1 \leq i \leq n$ and $x \in U_i \cap A$, we claim that $|R_i(y) - x| \leq \varepsilon r$ for any $y \in (U_i \backslash A) \cap B_{\varepsilon r_1}(x)$. Indeed, if $z \in U_i \cap B_{\varepsilon r_1}(x) \cap \partial A$ then $R_i(z) = z$ and therefore

$$|R_i(y) - x| \leq |R_i(y) - R_i(z)| + |z - x| \leq L^2|y - z| + |z - x| \leq (2L^2 + 1)\varepsilon r_1.$$

Thus, $R_i((U_i \backslash A) \cap B_{\varepsilon r_1}(x)) \subset U_i \cap A \cap B_{\varepsilon r}(x)$, and, after the change of variable $y' = R_i(y)$, we obtain

$$\int_{U_i \cap A} \int_{(U_i \backslash A) \cap B_{\varepsilon r_1}(x)} \left| \frac{u_i(y) - u_i(x)}{\varepsilon} \right|^p dy \, dx$$

$$\leq L^{2d} \int_{U_i \cap A} \int_{U_i \cap A \cap B_{\varepsilon r}(x)} \left| \frac{u(y') - u(x)}{\varepsilon} \right|^p dy' \, dx,$$

that controls the term in (4.4). Applying the same argument, we obtain an analogous estimate for the integral in (4.5). By the changes of variable $y' = R_i(y)$ and $x' = R_i(x)$ we also get

$$\int_{U_i \backslash A} \int_{(U_i \backslash A) \cap B_{\varepsilon r_1}(x)} \left| \frac{u_i(y) - u_i(x)}{\varepsilon} \right|^p dy \, dx$$

$$\leq L^{4d} \int_{U_j \cap A} \int_{U_i \cap A \cap B_{\varepsilon r}(x)} \left| \frac{u(y) - u(x)}{\varepsilon} \right|^p dy \, dx$$

and the integral in (4.6) is controlled as well. Hence (4.2) holds. Set

$$\varepsilon_0 = \varepsilon_0(r, A) = \frac{1}{r_1} \min_{0 \leq i \leq n} \, \text{dist}(V_i, U_i^c) > 0.$$

For every $\varepsilon < \varepsilon_0$ and $0 \leq i \leq n$, by (4.2) and (4.1), summing and subtracting $\varphi_i(x + \varepsilon\xi)u_i(x)$ we get

$$G_\varepsilon^{r_1}(\tilde{u}_i, \tilde{A}) \leq 2^{p-1} \int_{B_{r_1}} \int_{(U_i)_\varepsilon(\xi)} \left| \frac{u_i(x + \varepsilon\xi) - u_i(x)}{\varepsilon} \right|^p dx \, d\xi$$

$$+ 2^{p-1} \int_{B_{r_1}} \int_{(U_i)_\varepsilon(\xi)} |u_i(x)|^p \left| \frac{\varphi_i(x + \varepsilon\xi) - \varphi_i(x)}{\varepsilon} \right|^p dx \, d\xi$$

$$\leq C \left(G_\varepsilon^r(u, U_i \cap A) + \|u\|_{L^p(U_i \cap A; \mathbb{R}^m)}^p \right).$$

(4.7)

Hence, for every $\varepsilon < \varepsilon_0$ from (4.7) we obtain

$$G_\varepsilon^{r_1}(\tilde{u}, \tilde{A}) \leq \sum_{i=0}^n G_\varepsilon^{r_1}(\tilde{u}_i, \tilde{A}) \leq C \left(G_\varepsilon^r(u, A) + \|u\|_{L^p(A; \mathbb{R}^m)}^p \right).$$

Eventually, we notice that the map $E : u \mapsto \tilde{u}$ is linear by definition of $\tilde{u}$ which concludes the proof.   □

Note that, the theorem provides an extension to the whole space $\mathbb{R}^d$. Indeed, by construction we have that for every $u \in L^p(A; \mathbb{R}^m)$, $Eu \in L^p(\mathbb{R}^d; \mathbb{R}^m)$ with $\text{supp}(Eu) \Subset \tilde{A}$ and $G_\varepsilon^{r_1}(Eu, \tilde{A}) = G_\varepsilon^{r_1}(Eu, \mathbb{R}^d)$ for every $\varepsilon < \varepsilon_0$.

4.2 Control of Long-Range Interactions with Short-Range Interactions

With the following key result we show that interactions of any admissible range can be suitably controlled by the short-range energy G_ε^r.

Lemma 4.1 *For every $r > 0$ there exists a positive constant C such that, for any open set $E \subset \Omega$ and for every $\xi \in \mathbb{R}^d$, $\varepsilon > 0$ such that*

$$\varepsilon r < \text{dist}(E + (0, \varepsilon)\xi, \mathbb{R}^d \setminus \Omega)$$

(4.8)

and $u \in L^p(\Omega; \mathbb{R}^m)$, there holds

$$\int_E \left| \frac{u(x + \varepsilon\xi) - u(x)}{\varepsilon} \right|^p dx \leq C(|\xi|^p + 1) G_\varepsilon^r(u, E_{\varepsilon, \xi})$$

with $E_{\varepsilon, \xi} = E + (0, \varepsilon)\xi + B_{\varepsilon r}$ and $(0, \varepsilon)\xi = \{s\xi : s \in (0, \varepsilon)\}$.

Proof First, note that condition (4.8) implies that $E_{\varepsilon,\xi} \subset \Omega$ and thus the terms in the inequality above are well defined. For notational reasons we set $\varepsilon' = \varepsilon r/\sqrt{d+3}$. Let $R_\xi \in SO(d)$ be a rotation matrix such that $R_\xi e_1 = \xi/|\xi|$. We introduce the lattice $\mathcal{L}_\varepsilon := \{R_\xi i : i \in \varepsilon'\mathbb{Z}^d\}$ and, for any $j \in \mathcal{L}_\varepsilon$ define $Q_\varepsilon^j := j + R_\xi(-\varepsilon'/2, \varepsilon'/2)^d$ and set

$$\tilde{E}_{\varepsilon,\xi} := \bigcup_{j\in\mathcal{L}_\varepsilon} \{Q_\varepsilon^j : Q_\varepsilon^j \cap (E + (0,\varepsilon)\xi) \neq \emptyset\}.$$

Let $k = \lceil \varepsilon|\xi|/\varepsilon'\rceil + 1$, and, for any $j_0 \in \mathcal{L}_\varepsilon$ and $0 \le l \le k-1$, define $j_l = j_0 + l\varepsilon'\frac{\xi}{|\xi|}$. Denote by x_l any point in $Q_\varepsilon^{j_l}$ and, for the sake of simplicity of notation, $x_k = x_0 + \varepsilon\xi$ and $Q_\varepsilon^{j_k} = Q_\varepsilon^{j_0} + \varepsilon\xi$.

Using the inequality

$$\left|\frac{u(x_k) - u(x_0)}{\varepsilon}\right|^p \le k^{p-1} \sum_{l=1}^{k} \left|\frac{u(x_l) - u(x_{l-1})}{\varepsilon}\right|^p$$

and integrating in every variable we get

$$\int_{Q_\varepsilon^{j_0}} \left|\frac{u(x_0 + \varepsilon\xi) - u(x_0)}{\varepsilon}\right|^p dx_0$$

$$\le \frac{k^{p-1}}{(\varepsilon')^d} \sum_{l=1}^{k} \int_{Q_\varepsilon^{j_{l-1}}} \int_{Q_\varepsilon^{j_l}} \left|\frac{u(x_l) - u(x_{l-1})}{\varepsilon}\right|^p dx_l\, dx_{l-1}$$

$$\le \frac{k^{p-1}}{(\varepsilon')^d} \sum_{l=1}^{k} \int_{Q_\varepsilon^{j_{l-1}}} \int_{B_{\varepsilon r}(x_{l-1})} \left|\frac{u(y) - u(x_{l-1})}{\varepsilon}\right|^p dy\, dx_{l-1}.$$

Note that in the second line of the inequality above we have used that $Q_\varepsilon^{j_k} \subset Q_\varepsilon^{j_{k-2}} \cup Q_\varepsilon^{j_{k-1}}$ (see Fig. 4.1). Then, by the change of variable $y = x_{l-1} + \varepsilon\xi'$

$$\int_{Q_\varepsilon^{j_0}} \left|\frac{u(x_0 + \varepsilon\xi) - u(x_0)}{\varepsilon}\right|^p dx$$

$$\le \left(\frac{\varepsilon}{\varepsilon'}\right)^d k^{p-1} \sum_{l=1}^{k} \int_{B_r} \int_{Q_\varepsilon^{j_{l-1}}} \left|\frac{u(x_{l-1} + \varepsilon\xi') - u(x_{l-1})}{\varepsilon}\right|^p dx_{l-1}\, d\xi'$$

$$\le Ck^{p-1} \int_{B_r} \int_{Q_\varepsilon(j_0)} \left|\frac{u(x + \varepsilon\xi') - u(x)}{\varepsilon}\right|^p dx\, d\xi',$$

with $Q_\varepsilon(j_0) = \bigcup_{l=1}^{k-1} Q_\varepsilon^{j_l}$ and C a positive constant depending on r and d. Since the sets $\{Q_\varepsilon(j_0) : j_0 \in \mathcal{L}_\varepsilon\}$ overlap at most $k-1$ times, by summing over $j_0 \in \mathcal{L}_\varepsilon$

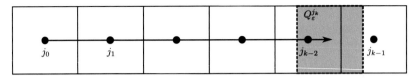

Fig. 4.1 The picture shows the cubes $Q_\varepsilon^{j_l}$ and the corresponding centers $j_l \in \mathcal{L}_\varepsilon$, $0 \le l \le k$. The arrow represents the vector $\varepsilon\xi$

such that $Q_\varepsilon^{j_0} \cap E \ne \emptyset$ we get that

$$\int_E \left| \frac{u(x + \varepsilon\xi) - u(x)}{\varepsilon} \right|^p dx \le C(|\xi|^p + 1) G_\varepsilon^r(u, \tilde{E}_{\varepsilon,\xi}).$$

Since $\text{dist}(E + (0, \varepsilon)\xi, \mathbb{R}^d \setminus \tilde{E}_{\varepsilon,\xi}) < \varepsilon r$ then $\tilde{E}_{\varepsilon,\xi} \subset E_{\varepsilon,\xi}$ and the result follows.   □

As a consequence of Lemma 4.1 and Theorem 4.1 we infer the following result.

Corollary 4.1 *For any open set $A \in \mathcal{A}^{\text{reg}}(\Omega)$ and $r > 0$ there exist two positive constants $C = C(r, A)$ and $\varepsilon_0 = \varepsilon_0(r, A)$ such that*

$$\int_{A_\varepsilon(\xi)} \left| \frac{u(x + \varepsilon\xi) - u(x)}{\varepsilon} \right|^p dx \le C(|\xi|^p + 1)\big(G_\varepsilon^r(u, A) + \|u\|^p_{L^p(A;\mathbb{R}^m)}\big),$$

for every $\xi \in \mathbb{R}^d$, $u \in L^p(A; \mathbb{R}^m)$, and $\varepsilon < \varepsilon_0$.

Remark 4.1   (short-range control) Let $\psi_{\varepsilon,2}$ be as in (2.5) and let $r, r' > 0$ be fixed. Then, Lemma 4.1 implies that there exist a positive constant $C = C(r, r')$ such that for every $u \in L^p(\Omega; \mathbb{R}^m)$ and $A \in \mathcal{A}(\Omega)$ there holds

$$G_\varepsilon[\psi_{\varepsilon,2}\,\chi_{\mathbb{R}^d \setminus B_{r'}}](u, A) \le C G_\varepsilon^r(u, A'),$$

for any $A' \in \mathcal{A}(\Omega)$ with $A' \supseteq A$, for every $\varepsilon < \frac{1}{2r}\text{dist}(A, (A')^c)$.

If moreover $\|\psi_{\varepsilon,2}\|_{L^1(\mathbb{R}^d)} \le C'$ then Lemma 4.1 provides a control also of short-range interactions; namely, for any given $r > 0$ there exists $C = C(r)$ such that $G_\varepsilon[\psi_{\varepsilon,2}](u, A) \le C G_\varepsilon^r(u, A')$ for every u, A, A' and ε as above.

4.3   Compactness in L^p Spaces

We now discuss the compactness in the strong L^p topology of sequences of functions with uniformly bounded energy and show that their limits are in the corresponding Sobolev space. In the scalar case and for bounded domains A, an

analogous result has been proved in [3, Theorems 1.2 and 1.3] for energies with
radially-symmetric kernels.

Theorem 4.2 *Let A be any open set of $\mathbb{R}^d$ with bounded Lipschitz boundary. Let*
$\{u_\varepsilon\}_\varepsilon \subset L^p(A; \mathbb{R}^m)$ *be such that for some $r > 0$*

$$\sup_{\varepsilon > 0} \left\{ \|u_\varepsilon\|_{L^p(A;\mathbb{R}^m)} + G_\varepsilon^r(u_\varepsilon, A) \right\} < +\infty.$$

*If A is unbounded, assume in addition that for any $\eta > 0$ there exists $r_\eta > 0$ such
that*

$$\sup_{\varepsilon > 0} \|u_\varepsilon\|_{L^p(A \setminus B_{r_\eta})} < \eta. \tag{4.9}$$

*Then, given $\varepsilon_j \to 0$, $\{u_{\varepsilon_j}\}_j$ is relatively compact in $L^p(A; \mathbb{R}^m)$ and every limit of a
converging subsequence is in $W^{1,p}(A; \mathbb{R}^m)$.*

Proof By Theorem 4.1, there exists $\tilde{A} \supseteq A$, $\tilde{r} > 0$ and $\tilde{u}_\varepsilon \in L^p(\tilde{A}; \mathbb{R}^m)$ such that
$\tilde{u}_\varepsilon = u_\varepsilon$ on A and

$$\|\tilde{u}_\varepsilon\|^p_{L^p(\tilde{A};\mathbb{R}^m)} + G_\varepsilon^{\tilde{r}}(\tilde{u}_\varepsilon, \tilde{A}) \leq C\|u_\varepsilon\|^p_{L^p(A;\mathbb{R}^m)} + CG_\varepsilon^r(u_\varepsilon, A).$$

Set $R = \text{dist}(A, \mathbb{R}^d \setminus \tilde{A})$ and let $\{\phi_\eta\}_{\eta < R}$ be a family of mollifiers; i.e., $\phi_\eta(x) =
\phi_1(x/\eta)/\eta^d$, where $\phi_1 \in C_c^\infty(\mathbb{R}^d)$, $0 \leq \phi_1 \leq 1$, $\text{supp}(\phi_1) \subset B_1$ and $\|\phi_1\|_{L^1(B_1)} =
1$. Note that the convolution product $\tilde{u}_\varepsilon * \phi_\eta(x)$ is well defined for every $x \in A$ and
by the standard properties of the convolution $\tilde{u}_\varepsilon * \phi_\eta \in C^\infty(A; \mathbb{R}^m)$ and

$$\|\tilde{u}_\varepsilon * \phi_\eta\|_{L^\infty(A;\mathbb{R}^m)} \leq \|\tilde{u}_\varepsilon\|_{L^p(\tilde{A};\mathbb{R}^m)}\|\phi_\eta\|_{L^{p'}(B_\eta)}$$

$$= \|\tilde{u}_\varepsilon\|_{L^p(\tilde{A};\mathbb{R}^m)}\|\phi_1\|_{L^{p'}(B_1)}\eta^{\frac{d(1-p')}{p'}}$$

$$\|\nabla(\tilde{u}_\varepsilon * \phi_\eta)\|_{L^\infty(A;\mathbb{R}^m)} \leq \|\tilde{u}_\varepsilon\|_{L^p(\tilde{A};\mathbb{R}^m)}\|\nabla\phi_\eta\|_{L^{p'}(B_\eta)} \tag{4.10}$$

$$= \|\tilde{u}_\varepsilon\|_{L^p(\tilde{A};\mathbb{R}^m)}\|\nabla\phi_1\|_{L^{p'}(B_1)}\eta^{\frac{d(1-p')}{p'}-1}.$$

Moreover, by Jensen's inequality we have

$$\|u_\varepsilon - \tilde{u}_\varepsilon * \phi_\eta\|^p_{L^p(A;\mathbb{R}^m)} = \int_A \left| u_\varepsilon(x) - \int_{B_\eta} \tilde{u}_\varepsilon(x-y)\phi_\eta(y)dy \right|^p dx$$

$$\leq \int_A \int_{B_\eta} |\tilde{u}_\varepsilon(x) - \tilde{u}_\varepsilon(x-y)|^p \phi_\eta(y)dy\,dx \tag{4.11}$$

$$= \int_{B_1} \int_A |\tilde{u}_\varepsilon(x) - \tilde{u}_\varepsilon(x+\eta\xi)|^p \phi_1(\xi)dx\,d\xi.$$

Taking $\eta = m\tilde{r}\varepsilon$ with $m \in \mathbb{N}$, by (4.11) and Jensen's inequality we get

$$\|u_\varepsilon - \tilde{u}_\varepsilon * \phi_{m\tilde{r}\varepsilon}\|^p_{L^p(A)}$$

$$\leq \int_{B_1} \int_A (m\varepsilon)^p \left| \sum_{l=0}^{m-1} \frac{\tilde{u}_\varepsilon(x + (l+1)\varepsilon\tilde{r}\xi) - \tilde{u}_\varepsilon(x + l\varepsilon\tilde{r}\xi)}{m\varepsilon} \right|^p dx\, d\xi$$

$$\leq m^{p-1}\varepsilon^p \sum_{l=0}^{m-1} \int_{B_1} \int_A \left| \frac{\tilde{u}_\varepsilon(x + (l+1)\varepsilon\tilde{r}\xi) - \tilde{u}_\varepsilon(x + l\varepsilon\tilde{r}\xi)}{\varepsilon} \right|^p dx\, d\xi$$

$$= m^{p-1}\varepsilon^p \sum_{l=0}^{m-1} \int_{B_1} \int_{A+l\varepsilon\tilde{r}\xi} \left| \frac{\tilde{u}_\varepsilon(x' + \varepsilon\tilde{r}\xi) - \tilde{u}_\varepsilon(x')}{\varepsilon} \right|^p dx'\, d\xi,$$

where in the last equality we have used the change of variable $x' = x + l\varepsilon\tilde{r}\xi$. Since $A + l\varepsilon\tilde{r}\xi \subset \tilde{A}_\varepsilon(\tilde{r}\xi)$ for every $0 \leq l \leq m-1$, we get

$$\|u_\varepsilon - \tilde{u}_\varepsilon * \phi_{m\tilde{r}\varepsilon}\|^p_{L^p(A;\mathbb{R}^m)} \leq (m\varepsilon)^p \int_{B_1} \int_{\tilde{A}_\varepsilon(\tilde{r}\xi)} \left| \frac{\tilde{u}_\varepsilon(x' + \varepsilon\tilde{r}\xi) - \tilde{u}_\varepsilon(x')}{\varepsilon} \right|^p dx'\, d\xi,$$

and, through the change of variable $\xi' = \tilde{r}\xi$, we obtain

$$\|u_\varepsilon - \tilde{u}_\varepsilon * \phi_{m\tilde{r}\varepsilon}\|^p_{L^p(A;\mathbb{R}^m)} \leq \frac{(m\varepsilon)^p}{\tilde{r}^d} G^{\tilde{r}}_\varepsilon(\tilde{u}_\varepsilon, \tilde{A}). \tag{4.12}$$

If A is unbounded, by applying Jensen's inequality to the L^p-norm of convolution $\tilde{u}_\varepsilon * \phi_\eta$ and (4.9), we can find $A_\eta \Subset A$ such that

$$\sup_{\varepsilon > 0} \|\tilde{u}_\varepsilon * \phi_\eta\|_{L^p(A \setminus A_\eta;\mathbb{R}^m)} < \eta. \tag{4.13}$$

Given $\varepsilon_j < \tilde{r}^{-1}\eta$ and setting $\eta_j = \lceil \eta/(\tilde{r}\varepsilon_j) \rceil \tilde{r}\varepsilon_j \geq \eta$, from (4.10) we get in particular

$$\|\tilde{u}_{\varepsilon_j} * \phi_{\eta_j}\|_{L^\infty(A_\eta;\mathbb{R}^m)} \leq C\eta^{\frac{d(1-p')}{p'}},$$

$$\|\nabla(\tilde{u}_{\varepsilon_j} * \phi_{\eta_j})\|_{L^\infty(A_\eta;\mathbb{R}^m)} \leq C\eta^{\frac{d(1-p')}{p'} - 1}.$$

Hence by Ascoli-Arzelá's Theorem, for every fixed η the sequence $\{\tilde{u}_{\varepsilon_j} * \phi_{\eta_j}\}_j$ is precompact in $C(A_\eta; \mathbb{R}^m)$ and therefore it is totally bounded, that is there exists a finite set of functions $\{g_k\}_{k=1}^L \subset C(A_\eta; \mathbb{R}^m)$ such that for every $j \in \mathbb{N}$

$$\|\tilde{u}_{\varepsilon_j} * \phi_{\eta_j} - g_k\|_{L^p(A_\eta;\mathbb{R}^m)} \leq |A_\eta|^{\frac{1}{p}} \|\tilde{u}_{\varepsilon_j} * \phi_{\eta_j} - g_k\|_{L^\infty(A_\eta;\mathbb{R}^m)} < \eta \tag{4.14}$$

for some $k \in \{1, \ldots, L\}$. So, by (4.12), (4.13) and (4.14), denoting by $\tilde{g}_k$ the extension of g_k which equals zero outside A_η, we have

$$
\begin{aligned}
\|u_{\varepsilon_j} - \tilde{g}_k\|_{L^p(A;\mathbb{R}^m)} &\le \|u_{\varepsilon_j} - \tilde{u}_{\varepsilon_j} * \phi_{\eta_j}\|_{L^p(A;\mathbb{R}^m)} + \|\tilde{u}_{\varepsilon_j} * \phi_{\eta_j} - \tilde{g}_k\|_{L^p(A;\mathbb{R}^m)} \\
&= \|u_{\varepsilon_j} - \tilde{u}_{\varepsilon_j} * \phi_{\eta_j}\|_{L^p(A;\mathbb{R}^m)} + \|\tilde{u}_{\varepsilon_j} * \phi_{\eta_j} - g_k\|_{L^p(A_\eta;\mathbb{R}^m)} \\
&\quad + \|\tilde{u}_{\varepsilon_j} * \phi_{\eta_j}\|_{L^p(A \setminus A_\eta)} \\
&\le \eta_j \frac{1}{\tilde{r}^{d/p+1}} G_{\varepsilon_j}^{\tilde{r}}(\tilde{u}_{\varepsilon_j}, A)^{\frac{1}{p}} + 2\eta \le C\eta;
\end{aligned}
$$

i.e., $\{u_{\varepsilon_j}\}_j$ is totally bounded and hence relatively compact in $L^p(A; \mathbb{R}^m)$. Finally Proposition 3.1 yields that every limit function is in $W^{1,p}(A; \mathbb{R}^m)$. □

Remark 4.2 From the previous theorem we deduce a compactness result on any open set A (without regularity assumptions on the boundary) with respect to the local L^p-topology. Namely, that every bounded sequence $\{u_\varepsilon\}$ with bounded G_ε^r-energy on A is precompact in $L^p(A'; \mathbb{R}^m)$ for every $A' \Subset A$. Indeed, it suffices to apply the previous theorem with a set A'' with Lipschitz boundary in the place of A, with $A' \Subset A'' \Subset A$.

4.4 Poincaré Inequalities

We complete this section with the asymptotic analogs of Poincaré inequalities. Similar results can be found e.g. in [3].

Proposition 4.1 (Poincaré Inequality) *Let $r > 0$ and let A be an open set in $\mathbb{R}^d$ such that $A \subseteq (a, b)v + \Pi_v$ for some $a, b \in \mathbb{R}$ and $v \in \mathbb{S}^{d-1}$, where $(a, b)v = \{sv : s \in (a, b)\}$ and $\Pi_v = \{x \in \mathbb{R}^d : \langle x, v \rangle = 0\}$. Then there exists a positive constant $C = C(r, A)$ such that*

$$
\int_A |u(x)|^p dx \le C G_\varepsilon^r(u, A)
$$

for every $\varepsilon > 0$ and for any $u \in L^p(A; \mathbb{R}^d)$ with $u(x) = 0$ for almost every $x \in A$ with $\operatorname{dist}(x, \mathbb{R}^d \setminus A) \le \varepsilon r$.

Proof We identify u with its extension to the whole $\mathbb{R}^d$ that equals zero outside A. It is not restrictive to consider $v = e_1$ and up to a change of variables we may assume $a = 0$ and $b = 1$. Set $r' := \frac{r}{\sqrt{d+3}}$ and for every $j \in \{0, \ldots, N := \lfloor 1/r'\varepsilon \rfloor\}$ and $l \in \mathbb{Z}^{d-1}$ denote by x_j^l an independent variable lying in

$$
Q_\varepsilon^{j,l} := (jr'\varepsilon, (j+1)r'\varepsilon) \times Q_\varepsilon^l, \quad \text{with } Q_\varepsilon^l := l + (0, r'\varepsilon)^{d-1}.
$$

By the boundary assumption, for any $0 \leq k \leq N$ we have

$$u(x_k^l) = \sum_{j=1}^{k} (u(x_j^l) - u(x_{j-1}^l)).$$

Thus, by Jensen's inequality we get

$$|u(x_k^l)|^p \leq k^{p-1} \sum_{j=1}^{k} |u(x_j^l) - u(x_{j-1}^l)|^p,$$

and, integrating in $x_0^l, x_1^l, \ldots, x_N^l$,

$$(r'\varepsilon)^{d(N-1)} \int_{Q_\varepsilon^{k,l}} |u(x_k^l)|^p dx_k^l$$

$$\leq (r'\varepsilon)^{d(N-2)} k^{p-1} \sum_{j=1}^{k} \int_{Q_\varepsilon^{j-1,l}} \int_{Q_\varepsilon^{j,l}} |u(x_j^l) - u(x_{j-1}^l)|^p dx_j^l dx_{j-1}^l.$$

Now, since $Q_\varepsilon^{j,l} \subset B_{r\varepsilon}(x_{j-1}^l)$, we have

$$\int_{Q_\varepsilon^{k,l}} |u(x_k^l)|^p dx_k^l \leq \frac{k^{p-1}}{(r'\varepsilon)^d} \sum_{j=1}^{k} \int_{Q_\varepsilon^{j-1,l}} \int_{B_{r\varepsilon}(x_{j-1}^l)} |u(y) - u(x_{j-1}^l)|^p dy dx_{j-1}^l$$

$$= \frac{k^{p-1}}{(r'\varepsilon)^d} \int_{(0,kr'\varepsilon) \times Q_\varepsilon^l} \int_{B_{r\varepsilon}(x)} |u(y) - u(x)|^p dy \, dx$$

$$\leq \frac{N^{p-1}}{(r')^d} \int_{(0,1) \times Q_\varepsilon^l} \int_{B_r} |u(x + \varepsilon\xi) - u(x)|^p d\xi dx.$$

By summing over $0 \leq k \leq N$ and $l \in \mathbb{Z}^{d-1}$ both the left- and the right-hand sides we get

$$\int_A |u(x)|^p dx \leq \frac{1}{(r')^{d+p}} \int_A \int_{B_r} \left| \frac{u(x + \varepsilon\xi) - u(x)}{\varepsilon} \right|^p d\xi dx$$

$$= \frac{1}{(r')^{d+p}} G_\varepsilon^r(u, A),$$

which proves the claim.   □

Proposition 4.2 (Poincaré-Wirtinger Inequality) *Let* $r > 0$ *and let* A *be a bounded connected open set of* $\mathbb{R}^d$ *with Lipschitz boundary. Then for every* *measurable set $E \subset A$ with $|E| > 0$ there exists a positive constant $C = C(r, A, E)$ such that*

$$\int_A |u(x) - u_E|^p dx \le C G_\varepsilon^r(u, A)$$

for any $u \in L^p(A; \mathbb{R}^m)$ and $\varepsilon \in (0, 1)$.

Proof We argue by contradiction. Suppose that for any positive integer j there exists $\varepsilon_j > 0$ and $u_j \in L^p(A; \mathbb{R}^m)$ such that

$$\int_A |u_j(x) - (u_j)_E|^p dx > j G_{\varepsilon_j}^r(u_j, A).$$

Thus, letting

$$\tilde{u}_j := \frac{u_j - (u_j)_E}{\|u_j - (u_j)_E\|_{L^p(A;\mathbb{R}^m)}},$$

we have $\|\tilde{u}_j\|_{L^p(A;\mathbb{R}^m)} \equiv 1$, $(\tilde{u}_j)_E \equiv 0$ and

$$G_{\varepsilon_j}^r(\tilde{u}_j, A) < \frac{1}{j}. \tag{4.15}$$

We may assume, up to passing to a subsequence, that $\varepsilon_j \to \varepsilon_0 \in [0, 1]$. If $\varepsilon_0 = 0$, Theorem 4.2 yields that up to subsequences $\tilde{u}_j \to u$ in $L^p(A; \mathbb{R}^m)$ with $u \in W^{1,p}(A; \mathbb{R}^m)$, $\|u\|_{L^p(A;\mathbb{R}^m)} = 1$ and $u_E = 0$. By (4.15) and Proposition 3.1, we get

$$\int_{B_r} \int_A |\nabla u(x)\xi|^p dx\, d\xi = 0.$$

Hence, u is constant almost everywhere and we reach a contradiction. If otherwise $\varepsilon_0 > 0$, up to passing to a further subsequence, $\tilde{u}_j \rightharpoonup u$ weakly in $L^p(A; \mathbb{R}^m)$ and so

$$\frac{\tilde{u}_j(x + \varepsilon_j \xi) - \tilde{u}_j(x)}{\varepsilon_j} \rightharpoonup \frac{u(x + \varepsilon_0 \xi) - u(x)}{\varepsilon_0} \quad \text{weakly in } L^p(A; \mathbb{R}^m)$$

for any $\xi \in B_r$.

For every open set $A' \Subset A_{\varepsilon_0}(\xi)$, Fatou's Lemma yields

$$0 = \liminf_{j \to \infty} G_{\varepsilon_j}^r(\tilde{u}_j, A)$$

$$\ge \int_{B_r} \liminf_{j \to \infty} \int_{A'} \left| \frac{\tilde{u}_j(x + \varepsilon_j \xi) - \tilde{u}_j(x)}{\varepsilon_j} \right|^p dx\, d\xi.$$

Hence, by the arbitrariness of A', we have that $G^r_{\varepsilon_0}(u, A) = 0$ and therefore u is constant almost everywhere on A, which again leads to a contradiction.   □

By means of the Poincaré-Wirtinger inequality we can improve Theorem 4.2. Indeed, we recover the compactness result by assuming the uniform boundedness of the mean values of u_ε.

Corollary 4.2 *Let A be any open bounded set of $\mathbb{R}^d$ with Lipschitz boundary. Let $E \subset A$ be a measurable set with $|E| > 0$. Let $\{u_\varepsilon\}_\varepsilon \subset L^p(A; \mathbb{R}^m)$ be such that for some $r > 0$*

$$\sup_{\varepsilon > 0} \left\{ |(u_\varepsilon)_E| + G^r_\varepsilon(u, A) \right\} < +\infty.$$

Then the same conclusion of Theorem 4.2 holds.

Proof It is sufficient to prove that $\|u_\varepsilon\|_{L^p(A;\mathbb{R}^m)}$ is uniformly bounded. By Proposition 4.2 we have

$$\|u_\varepsilon\|^p_{L^p(A;\mathbb{R}^m)} \leq 2^{p-1} \|u_\varepsilon - (u_\varepsilon)_E\|^p_{L^p(A;\mathbb{R}^m)} + 2^{p-1}|A||(u_\varepsilon)_E|^p \leq C(G^r_\varepsilon(u_\varepsilon, A) + 1)$$

for some constant C depending on A and r and the claim follows.   □

References

1. Di Nezza, E., Palatucci, G., Valdinoci, E.: Hitchhiker's guide to the fractional Sobolev spaces. Bull. Sci. Math. **136**, 521–573 (2012)
2. Leoni, G.: A First Course in Sobolev Spaces. Graduate Students in Mathematics, vol. 181, 2nd edn. American Mathematical Society, Providence (2017)
3. Ponce, A.C.: An estimate in the spirit of Poincaré's inequality. J. Eur. Math. Soc. **6**, 1–15 (2004)

Chapter 5
A Compactness and Integral-Representation Result

Abstract The main result of this chapter is a compactness and integral-representation result for the Γ-limits of the families $\{F_\varepsilon(\cdot, A)\}_\varepsilon$, which we can obtain through a convolution version of the localization method of Γ-convergence. A key point is that it is possible to limit the analysis to finite-range convolutions through a truncation argument.

Keywords Direct method of Γ-convergence · Compactness · Integral representation · Truncations · Dirichlet boundary conditions · Boundary-value problems · Euler-Lagrange equations

5.1 The Integral-Representation Theorem

We now prove that from any family of functionals $\{F_\varepsilon\}_\varepsilon$ we can extract a Γ-converging sequence, whose limit can be represented as a local integral functional. Its proof will follow a strategy devised by De Giorgi and described, e.g., in [2, Chapter 9]. It consists in considering the localized functionals

$$F_\varepsilon(u, A) := \int_{\mathbb{R}^d} \int_{A_\varepsilon(\xi)} f_\varepsilon\left(x, \xi, \frac{u(x + \varepsilon\xi) - u(x)}{\varepsilon}\right) dx \, d\xi,$$

already introduced in (2.3), and in proving properties of their Γ-limit both in terms of the variable u and the set variable A. The result is stated as follows.

Theorem 5.1 (Compactness and Integral Representation) *Given Ω a bounded open set with Lipschitz boundary, let F_ε, $F_\varepsilon(\cdot, \cdot)$ be defined by (2.1) and (2.3), respectively, and let assumptions (H0'), (H1) and (H2) be satisfied. Then, for every $\varepsilon_j \to 0$ there exists a subsequence $\{\varepsilon_{j_k}\} \subset \{\varepsilon_j\}$ and a Carathédory function $f_0 : \Omega \times \mathbb{R}^{m \times d} \to [0, +\infty)$ which is quasiconvex in the second variable and satisfies the growth condition*

$$C_0(|M|^p - 1) \leq f_0(x, M) \leq C_1(|M|^p + 1) \tag{5.1}$$

© The Author(s), under exclusive license to Springer Nature Singapore Pte Ltd. 2023
R. Alicandro et al., *A Variational Theory of Convolution-Type Functionals*,
SpringerBriefs on PDEs and Data Science,
https://doi.org/10.1007/978-981-99-0685-7_5

for almost every $x \in \Omega$ and every $M \in \mathbb{R}^{m \times d}$, such that

$$\Gamma(L^p)\text{-}\lim_{k \to +\infty} F_{\varepsilon_{j_k}}(u, A) = \begin{cases} \int_A f_0(x, \nabla u(x))dx & \text{if } u \in W^{1,p}(A; \mathbb{R}^m) \\ +\infty & \text{otherwise,} \end{cases} \tag{5.2}$$

for every $A \in \mathcal{A}^{\mathrm{reg}}(\Omega)$.

In order to prove Theorem 5.1 we will show that, up to subsequences, the Γ-limit $F(u, A)$ exists for all $A \in \mathcal{A}^{\mathrm{reg}}(\Omega)$, and that we can apply the following result.

Theorem 5.2 (Theorem 9.1 [2]) *Let Ω be a bounded open set, and let F : $W^{1,p}(\Omega; \mathbb{R}^m) \times \mathcal{A}(\Omega) \to [0, +\infty)$ satisfy:*

(i) *for any $A \in \mathcal{A}(\Omega)$ $F(u, A) = F(v, A)$ if $u = v$ almost everywhere on A;*
(ii) *for any $u \in W^{1,p}(\Omega; \mathbb{R}^m)$ the set function $F(u, \cdot)$ is a restriction of a Borel measure on $\mathcal{A}(\Omega)$;*
(iii) *there exists a constant $C > 0$ and $a \in L^1(\Omega)$ such that*

$$F(u, A) \leq C \int_A (a(x) + |\nabla u(x)|^p)\, dx;$$

(iv) *$F(u + z, A) = F(u, A)$ for any $u \in W^{1,p}(\Omega; \mathbb{R}^m)$, $z \in \mathbb{R}^m$ and $A \in \mathcal{A}(\Omega)$;*
(v) *$F(\cdot, A)$ is weakly lower semicontinuous for any $A \in \mathcal{A}(\Omega)$.*

Then there exists a Carathéodory function $f : \Omega \times \mathbb{R}^{m \times d} \to [0, +\infty)$, quasiconvex in the second variable, with $0 \leq f(x, M) \leq c(a(x) + |M|^p)$ for almost every $x \in A$ and every $M \in \mathbb{R}^{m \times d}$, such that

$$F(u, A) = \int_A f(x, \nabla u(x))dx$$

for every $A \in \mathcal{A}(\Omega)$ and $u \in W^{1,p}(\Omega; \mathbb{R}^d)$.

We postpone the proof of Theorem 5.1 to the end of the section as it will be a direct consequence of some propositions that show that the limit functionals satisfy the hypotheses of Theorem 5.2.

5.2   Truncated-Range Functionals

In this subsection we introduce the ' truncated functionals' obtained by limiting the range of interaction in (2.1) to a fixed threshold $T > 0$. We will show that, to some extent, the limit as $T \to +\infty$ and the Γ-limit as $\varepsilon \to 0$ commute. This result will allow to limit our analysis to truncated functionals in the proofs of the results of the following sections, in particular in that of Theorem 5.1, leading to significant simplifications.

Definition 5.1 (The *Truncated Functionals* F_ε^T) For any $A \in \mathcal{A}(\Omega)$ and $T > 0$ the functional $F_\varepsilon^T(\cdot, A) : L^p(A; \mathbb{R}^m) \to [0, +\infty]$ is defined by

$$F_\varepsilon^T(u, A) := \int_{B_T} \int_{A_\varepsilon(\xi)} f_\varepsilon\left(x, \xi, \frac{u(x + \varepsilon\xi) - u(x)}{\varepsilon}\right) dx \, d\xi . \tag{5.3}$$

Note that Proposition 3.3 clearly applies also to the truncated functionals $F^T(\cdot, A)$, since they comply with all the hypotheses of Sect. 2.3. In what follows we use the notation

$$F'^{,T}(u, A) := \Gamma\text{-}\liminf_{\varepsilon \to 0} F_\varepsilon^T(u, A), \quad F''^{,T}(u, A) := \Gamma\text{-}\limsup_{\varepsilon \to 0} F_\varepsilon^T(u, A).$$

Lemma 5.1 *Let $F_\varepsilon(\cdot, A)$ and $F_\varepsilon^T(\cdot, A)$ be defined by (2.3) and (5.3), respectively, and let assumptions (H0)–(H2) be satisfied. Then for every $A \in \mathcal{A}^{\mathrm{reg}}(\Omega)$ and $u \in L^p(\Omega; \mathbb{R}^m)$*

$$F'(u, A) = \lim_{T \to +\infty} F'^{,T}(u, A) \quad \text{and} \quad F''(u, A) = \lim_{T \to +\infty} F''^{,T}(u, A).$$

In particular, given $T_j \to +\infty$ such that $F_\varepsilon^{T_j}(\cdot, A)$ Γ-converge to $F^{T_j}(\cdot, A)$ as $\varepsilon \to 0$ for every $j \in \mathbb{N}$,

$$\Gamma\text{-}\lim_{\varepsilon \to 0} F_\varepsilon(u, A) = \lim_{j \to +\infty} F^{T_j}(u, A)$$

for every $u \in L^p(\Omega; \mathbb{R}^m)$.

Proof Note first that, since $F_\varepsilon^T(u, A) \leq F_\varepsilon(u, A)$ for every $u \in L^p(\Omega; \mathbb{R}^m)$ and $A \in \mathcal{A}(\Omega)$, one inequality in the statement is trivial. Thanks to Proposition 3.3, it is sufficient to prove the opposite inequality for every $u \in W^{1,p}(A; \mathbb{R}^m)$. Hence, let $u_\varepsilon \to u$ in $L^p(\Omega; \mathbb{R}^m)$. Without loss of generality we may assume that $F_\varepsilon(u_\varepsilon, A)$ is uniformly bounded. We have that

$$F_\varepsilon(u_\varepsilon, A) = F_\varepsilon^T(u_\varepsilon, A) + \int_{B_T^c} \int_{A_\varepsilon(\xi)} f_\varepsilon\left(x, \xi, \frac{u_\varepsilon(x + \varepsilon\xi) - u_\varepsilon(x)}{\varepsilon}\right) dx \, d\xi,$$

and from assumption (H1) we get

$$F_\varepsilon(u_\varepsilon, A) \leq F_\varepsilon^T(u_\varepsilon, A)$$

$$+ \int_{B_T^c} \int_{A_\varepsilon(\xi)} \left(\psi_{\varepsilon,2}(\xi) \left|\frac{u_\varepsilon(x + \varepsilon\xi) - u_\varepsilon(x)}{\varepsilon}\right|^p + \rho_{\varepsilon,2}(\xi)\right) dx \, d\xi. \tag{5.4}$$

Thanks to Corollary 4.1, for ε small enough

$$\int_{B_T^c} \int_{A_\varepsilon(\xi)} \left(\psi_{\varepsilon,2}(\xi) \left| \frac{u_\varepsilon(x + \varepsilon\xi) - u_\varepsilon(x)}{\varepsilon} \right|^p + \rho_{\varepsilon,2}(\xi) \right) dx \, d\xi$$

$$\leq \int_{B_T^c} \left(\psi_{\varepsilon,2}(\xi) C(|\xi|^p + 1) \left(G_\varepsilon^{r_0}(u_\varepsilon, A) + \|u_\varepsilon\|_{L^p(A;\mathbb{R}^m)}^p \right) + \rho_{\varepsilon,2}(\xi)|A| \right) d\xi \, .$$

$$(5.5)$$

Since by (2.9) $G_\varepsilon^{r_0}(u_\varepsilon, A)$ is bounded, by (5.4), (5.5) and (H2) we have

$$F_\varepsilon(u_\varepsilon, A) \leq F_\varepsilon^T(u_\varepsilon, A) + C\delta + o(1)$$

for every $T > r_\delta$, where r_δ is chosen as in (H2), and the thesis follows letting first ε and then δ tend to 0. □

5.3   Fundamental Estimates

In what follows, with a slight abuse of notation, $F''(\cdot, \cdot)$ will denote the Γ-lim sup of both the family of functionals $\{F_\varepsilon\}_\varepsilon$ and the sequence $\{F_{\varepsilon_j}\}_j$ for any $\varepsilon_j \to 0$.

A crucial step in order to apply Theorem 5.2 is provided by the following two propositions.

Proposition 5.1 (Subadditivity) *Let $F_\varepsilon(\cdot, \cdot)$ be defined by (2.3) and assume that (H0)–(H2) hold. Let $A, B \in \mathcal{A}(\Omega)$ and let $A', B' \in \mathcal{A}(\Omega)$ with $A' \Subset A$ and $B' \Subset B$. Then*

$$F''(u, A' \cup B') \leq F''(u, A) + F''(u, B) \tag{5.6}$$

for every $u \in L^p(\Omega; \mathbb{R}^m)$.

Proof Without loss of generality we may suppose that $F''(u, A)$ and $F''(u, B)$ are finite. Moreover, since $F''(u, \cdot)$ is an increasing set function, we may assume that $A', B' \in \mathcal{A}^{\text{reg}}(\Omega)$. Let $u_\varepsilon, v_\varepsilon$ both converge to u in $L^p(\Omega; \mathbb{R}^d)$ and be such that

$$\lim_{\varepsilon \to 0} F_\varepsilon(u_\varepsilon, A) = F''(u, A), \quad \lim_{\varepsilon \to 0} F_\varepsilon(v_\varepsilon, B) = F''(u, B).$$

Note that, by (2.8) and (2.9), $G_\varepsilon[\psi_{\varepsilon,1}](u_\varepsilon, A)$, $G_\varepsilon[\psi_{\varepsilon,1}](v_\varepsilon, B)$, $G_\varepsilon^{r_0}(u_\varepsilon, A)$ and $G_\varepsilon^{r_0}(v_\varepsilon, B)$ are uniformly bounded. Let $R := \text{dist}(A', \mathbb{R}^d \backslash A)$ and, fixed $N \in \mathbb{N}$, set

$$A_i = \{x \in A : \text{dist}(x, A') < iR/N\}, \ 1 \leq i \leq N.$$

Let φ^i be a cut-off function between A_i and A_{i+1}, with $|\nabla\varphi^i| \leq 2N/R$, and set $w_\varepsilon^i := u_\varepsilon\varphi^i + v_\varepsilon(1 - \varphi^i)$. Note that $w_\varepsilon^i \to u$ in $L^p(\Omega; \mathbb{R}^m)$ for any $i \in \{1, \ldots, N\}$. By adding and subtracting $\varphi^i(x)u_\varepsilon(x + \varepsilon\xi) + (1 - \varphi^i(x))v_\varepsilon(x + \varepsilon\xi)$ we have

$$w_\varepsilon^i(x + \varepsilon\xi) - w_\varepsilon^i(x) = \varphi^i(x)(u_\varepsilon(x + \varepsilon\xi) - u_\varepsilon(x))$$

$$+ (1 - \varphi^i(x))(v_\varepsilon(x + \varepsilon\xi) - v_\varepsilon(x))$$

$$+ (\varphi^i(x + \varepsilon\xi) - \varphi^i(x))(u_\varepsilon(x + \varepsilon\xi) - v_\varepsilon(x + \varepsilon\xi)).$$

$$(5.7)$$

Note that

$$w_\varepsilon^i(x + \varepsilon\xi) - w_\varepsilon^i(x) = \begin{cases} u_\varepsilon(x + \varepsilon\xi) - u_\varepsilon(x), & \text{if } x \in (A_i)_\varepsilon(\xi) \\ v_\varepsilon(x + \varepsilon\xi) - v_\varepsilon(x), & \text{if } x \in (\Omega\backslash\overline{A}_{i+1})_\varepsilon(\xi), \end{cases} \qquad (5.8)$$

while, having set

$$S_{\varepsilon,\xi}^i := (A' \cup B')_\varepsilon(\xi) \setminus \big((A_i)_\varepsilon(\xi) \cup (\Omega\backslash\overline{A}_{i+1})_\varepsilon(\xi)\big),$$

by (5.7) and Jensen's inequality, using the notation $D_\xi^\varepsilon g(x) = (g(x + \varepsilon\xi) - g(x))/\varepsilon$ for the sake of brevity, we get

$$|D_\xi^\varepsilon w_\varepsilon^i(x)|^p \leq 2^{p-1}\varphi^i(x)|D_\xi^\varepsilon u_\varepsilon(x)|^p + 2^{p-1}(1 - \varphi^i(x))|D_\xi^\varepsilon v_\varepsilon(x)|^p$$

$$+ 2^{p-1}|D_\xi^\varepsilon\varphi^i(x)|^p|u_\varepsilon(x + \varepsilon\xi) - v_\varepsilon(x + \varepsilon\xi)|^p$$

$$\leq 2^{p-1}\big(|D_\xi^\varepsilon u_\varepsilon(x)|^p + |D_\xi^\varepsilon v_\varepsilon(x)|^p\big) \qquad (5.9)$$

$$+ 2^{p-1}\left(\frac{2N}{R}\right)^p|\xi|^p|u_\varepsilon(x + \varepsilon\xi) - v_\varepsilon(x + \varepsilon\xi)|^p$$

for every $x \in S_{\varepsilon,\xi}^i$ (see Fig. 5.1).

Now, we consider the truncated functionals F_ε^T introduced in Definition 5.1. By (5.8), (5.9) and assumption (H1) we have

$$F_\varepsilon^T(w_\varepsilon^i, A' \cup B')$$

$$= F_\varepsilon^T(u_\varepsilon, A_i \cap (A' \cup B')) + F_\varepsilon^T(v_\varepsilon, (\Omega\backslash\overline{A}_{i+1}) \cap B')$$

$$+ \int_{B_T} \int_{S_{\varepsilon,\xi}^i} f_\varepsilon(x, \xi, D_\xi^\varepsilon w_\varepsilon^i(x))\, dx\, d\xi$$

$$\leq F_\varepsilon(u_\varepsilon, A) + F_\varepsilon(v_\varepsilon, B)$$

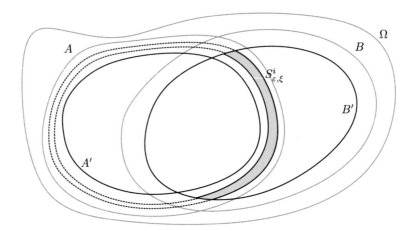

Fig. 5.1 The shaded region represents the set $S^i_{\varepsilon,\xi}$. The dashed lines are the boundaries of A_i and A_{i+1}. In this picture $\xi = e_1$

$$+ C \int_{B_T} \int_{S^i_{\varepsilon,\xi}} \left(\psi_{\varepsilon,2}(\xi)(|D^\varepsilon_\xi u_\varepsilon(x)|^p + |D^\varepsilon_\xi v_\varepsilon(x)|^p) + \rho_{\varepsilon,2}(\xi) \right) dx \, d\xi$$

$$(5.10)$$

$$+ C N^p \int_{B_T} \psi_{\varepsilon,2}(\xi) \int_{S^i_{\varepsilon,\xi}} |\xi|^p |u_\varepsilon(x + \varepsilon\xi) - v_\varepsilon(x + \varepsilon\xi)|^p dx \, d\xi. \qquad (5.11)$$

The integral in (5.11) can be estimated from above uniformly in T as follows

$$\int_{B_T} \psi_{\varepsilon,2}(\xi) \int_{S^i_{\varepsilon,\xi}} |\xi|^p |u_\varepsilon(x + \varepsilon\xi) - v_\varepsilon(x + \varepsilon\xi)|^p dx \, d\xi$$

$$\leq \int_{\mathbb{R}^d} \psi_{\varepsilon,2}(\xi) |\xi|^p \|u_\varepsilon - v_\varepsilon\|^p_{L^p(\Omega;\mathbb{R}^m)} d\xi$$

$$\leq C \|u_\varepsilon - v_\varepsilon\|^p_{L^p(\Omega;\mathbb{R}^m)};$$

hence it tends to zero as $\varepsilon \to 0$, since u_ε and v_ε both converge to u in $L^p(\Omega; \mathbb{R}^m)$. Note that, if $|\xi| < T$ then $S^i_{\varepsilon,\xi} \subset (A_{i+1}\backslash \overline{A_i} + (-\varepsilon, \varepsilon)\xi) \cap B'$; hence,

$$\bigcup_{i=1}^{N-4} S^i_{\varepsilon,\xi} \subset (A_{N-2}\backslash \overline{A'}) \cap B'. \qquad (5.12)$$

Moreover, the sets $\{S_{\varepsilon,\xi}^i\}_i$ intersect at most pairwise for ε small enough. Thus, from (5.12) for ε small enough we get

$$\sum_{i=1}^{N-4} \int_{B_T} \psi_{\varepsilon,2}(\xi) \int_{S_{\varepsilon,\xi}^i} \left(|D_\xi^\varepsilon u_\varepsilon(x)|^p + |D_\xi^\varepsilon v_\varepsilon(x)|^p\right) dx\, d\xi$$

$$\leq 2 \int_{B_T} \psi_{\varepsilon,2}(\xi) \int_{A_{N-2}\cap B'} \left(|D_\xi^\varepsilon u_\varepsilon(x)|^p + |D_\xi^\varepsilon v_\varepsilon(x)|^p\right) dx\, d\xi.$$

Here, we proceed in two different ways according to the dichotomy of (2.7). If $\limsup_{\varepsilon\to0} \int_{\mathbb{R}^d} \psi_{\varepsilon,2}(\xi)d\xi < +\infty$ then we apply Lemma 4.1 on u_ε and v_ε with $E = A_{N-2}\cap B'$ for any $|\xi| < T$, so that $E_{\varepsilon,\xi} \subset A_{N-1}\cap(B' + B_{\varepsilon(r_0+T)})$. Thus for ε small enough $E_{\varepsilon,\xi} \subset A\cap B$ and we get

$$\sum_{i=1}^{N-4} \int_{B_T} \psi_{\varepsilon,2}(\xi) \int_{S_{\varepsilon,\xi}^i} \left(|D_\xi^\varepsilon u_\varepsilon(x)|^p + |D_\xi^\varepsilon v_\varepsilon(x)|^p\right) dx\, d\xi$$

$$\leq (G_\varepsilon^{r_0}(u_\varepsilon, A) + G_\varepsilon^{r_0}(v_\varepsilon, B)) \int_{B_T} C\psi_{\varepsilon,2}(\xi)(|\xi|^p + 1)d\xi \leq C',$$

for some $C' > 0$. If instead $\limsup_{\varepsilon\to0} \int_{\mathbb{R}^d} \psi_{\varepsilon,2}(\xi)d\xi = +\infty$ by (2.7) and applying Lemma 4.1 we obtain

$$\sum_{i=1}^{N-4} \int_{B_T} \psi_{\varepsilon,2}(\xi) \int_{S_{\varepsilon,\xi}^i} \left(|D_\xi^\varepsilon u_\varepsilon(x)|^p + |D_\xi^\varepsilon v_\varepsilon(x)|^p\right) dx\, d\xi$$

$$\leq 2c_1\left(G_\varepsilon[\psi_{\varepsilon,1}](u_\varepsilon, A) + G_\varepsilon[\psi_{\varepsilon,1}](v_\varepsilon, B)\right)$$

$$+ (G_\varepsilon^{r_0}(u_\varepsilon, A) + G_\varepsilon^{r_0}(v_\varepsilon, B)) \int_{B_T\setminus B_{r_0}} C\psi_{\varepsilon,2}(\xi)(|\xi|^p + 1)d\xi \leq C',$$

where the last inequality is a consequence of (2.5) and the fact that the integration domain is far from the origin. We can choose an index $1 \leq k_\varepsilon \leq N - 4$ such that

$$\int_{B_T} \int_{S_{\varepsilon,\xi}^{k_\varepsilon}} (\psi_{\varepsilon,2}(\xi)(|D_\xi^\varepsilon u_\varepsilon(x)|^p + |D_\xi^\varepsilon v_\varepsilon(x)|^p) + \rho_{\varepsilon,2})\, dx\, d\xi \leq \frac{C}{N-4}, \quad (5.13)$$

uniformly in T, and

$$\Gamma\text{-}\limsup_{\varepsilon\to0} F_\varepsilon^T(u, A' \cup B') \leq \Gamma\text{-}\limsup_{\varepsilon\to0} F_\varepsilon^T(w_\varepsilon^{k_\varepsilon}, A' \cup B')$$

$$\leq F''(u, A) + F''(u, B) + \frac{C}{N-4}.$$

Letting first $N \to +\infty$ and then $T \to +\infty$, by Lemma 5.1 we get the conclusion.

$\square$

Remark 5.1 Reasoning as in Proposition 5.1 we can also derived the following inequality for the Γ-lim inf

$$F'(u, A' \cup B') \leq F'(u, A) + F''(u, B). \qquad (5.14)$$

Proposition 5.2 (Inner Regularity) *Let $F_\varepsilon(\cdot, \cdot)$ be defined by* (2.3) *and assume that* (H0)–(H2) *hold. Then*

$$\sup_{A' \Subset A} F''(u, A') = F''(u, A)$$

for every $A \in \mathcal{A}^{\mathrm{reg}}(\Omega)$ and $u \in L^p(\Omega; \mathbb{R}^m) \cap W^{1,p}(A; \mathbb{R}^m)$.

Proof Since $F''(u, \cdot)$ is an increasing set-function, it suffices to prove that

$$\sup_{A' \Subset A} F''(u, A') \geq F''(u, A).$$

To this end we argue as in the proof of Proposition 5.1. For any $\delta > 0$ let $A_\delta \Subset A$ be an open set such that

$$|A \setminus \overline{A_\delta}| + \|Du\|^p_{L^p(A \setminus \overline{A_\delta})} < \delta$$

and let $A' \in \mathcal{A}(\Omega)$ be such that $A_\delta \Subset A' \Subset A$. Let $u_\varepsilon, v_\varepsilon \in L^p(\Omega; \mathbb{R}^m)$ both converge to u in $L^p(\Omega; \mathbb{R}^m)$ and be such that

$$\lim_{\varepsilon \to 0} F_\varepsilon(u_\varepsilon, A') = F''(u, A'), \quad \lim_{\varepsilon \to 0} F_\varepsilon(v_\varepsilon, A \setminus \overline{A_\delta}) = F''(u, A \setminus \overline{A_\delta}).$$

Thus, by Proposition 3.3 we get

$$F_\varepsilon(v_\varepsilon, A \setminus \overline{A_\delta}) \leq F''(u, A \setminus \overline{A_\delta}) + o_\varepsilon(1) \leq C\delta + o_\varepsilon(1). \qquad (5.15)$$

Set $R := \mathrm{dist}(A_\delta, \Omega \setminus A')$, $A_i = \{x \in A : \mathrm{dist}(x, A_\delta) < iR/N\}$, and let φ_i and w_ε^i be defined as in the proof of Proposition 5.1. Set

$$S_{\varepsilon,\xi}^i := A_\varepsilon(\xi) \setminus ((A_i)_\varepsilon(\xi) \cup (A \setminus \overline{A_{i+1}})_\varepsilon(\xi)) \subset (A_{i+1} \setminus \overline{A_i} + (-\varepsilon, \varepsilon)\xi) \cap A_\varepsilon(\xi).$$

Consider first the finite-range interaction energies $F_\varepsilon^T(\cdot, \cdot)$. Reasoning as in the proof of Proposition 5.1, we can find $2 \le k_\varepsilon \le N - 4$ such that

$$F_\varepsilon^T(w_\varepsilon^{k_\varepsilon}, A)$$

$$= F_\varepsilon^T(u_\varepsilon, A_{k_\varepsilon}) + F_\varepsilon^T(v_\varepsilon, A \setminus \overline{A_{k_\varepsilon+1}}) + \int_{B_T} \int_{S_{\varepsilon,\xi}^{k_\varepsilon}} f_\varepsilon(x, \xi, D_\xi^\varepsilon w_\varepsilon^k(x)) dx \, d\xi$$

$$\le F_\varepsilon(u_\varepsilon, A') + o_\varepsilon(1) + \frac{C}{N-4} + C\delta.$$

Letting first $\varepsilon \to 0$ and then $N \to +\infty$ we get

$$F''^{,T}(u, A) \le F''(u, A') + C\delta \le \sup_{A' \Subset A} F''(u, A') + C\delta.$$

The result follows from the arbitrariness of δ and T and Lemma 5.1. $\qquad\square$

Remark 5.2 We can also prove the inner regularity of F'. Indeed, we may consider $u_\varepsilon, v_\varepsilon \in L^p(\Omega; \mathbb{R}^m)$ both converge to u in $L^p(\Omega; \mathbb{R}^m)$ and be such that

$$\liminf_{\varepsilon \to 0} F_\varepsilon(u_\varepsilon, A') \ge F'(u, A'), \quad \lim_{\varepsilon \to 0} F_\varepsilon(v_\varepsilon, A \setminus \overline{A_\delta}) = F''(u, A \setminus \overline{A_\delta}).$$

By (5.14) and reasoning as in Proposition 5.2 we get

$$\sup_{A' \Subset A} F'(u, A') = F'(u, A).$$

Note that, when dealing with local functionals, the fundamental estimates immediately implies the inner regularity of $F'(u, \cdot)$, $F''(u, \cdot)$ as done in [2]. In the non-local framework, it is important to be 'far' from $A' \cap \partial B$ because otherwise we can have interactions with points outside B that cannot be controlled by $G_\varepsilon^{r_0}(v_\varepsilon, B)$.

5.4 Proof of the Integral-Representation Theorem

This section is devoted to the proof of Theorem 5.1. We divide the proof in two steps, dealing first with the case in which (H0) holds and with the general case.

Step 1 Assume that (H0) holds. The compactness properties of Γ-convergence, Proposition 5.2, Remark 5.2 and [2, Theorem 10.3] yield the existence of a subsequence (ε_{j_k}) such that the Γ-limit

$$\Gamma(L^p)\text{-}\lim_{k \to \infty} F_{\varepsilon_{j_k}}(u, A) =: F(u, A)$$

exists for any $(u, A) \in L^p(\Omega; \mathbb{R}^m) \times \mathcal{A}^{\text{reg}}(\Omega)$. By Proposition 3.3, $F(u, A) = +\infty$ if and only if $u \notin W^{1,p}(A; \mathbb{R}^m)$. We now introduce the inner-regular extension of $F(u, \cdot)$ on the whole family $\mathcal{A}(\Omega)$ defined by

$$\tilde{F}(u, A) := \sup\{F(u, A') : A' \in \mathcal{A}^{\text{reg}}(\Omega), \ A' \Subset A\}.$$

Since, by Proposition 5.2, $\tilde{F}(u, A) = F(u, A)$ for any $u \in W^{1,p}(A; \mathbb{R}^m)$ and $A \in \mathcal{A}^{\text{reg}}(\Omega)$, it remains to check that $\tilde{F}$ satisfies all the hypotheses of Theorem 5.2. $\tilde{F}(u, \cdot)$ is clearly increasing. By Remark 2.2, hypothesis (i) trivially holds. Proposition 3.3 yields (iii). Since $F_\varepsilon(\cdot, A)$ depends only on incremental ratios, it is translation invariant and so does $\tilde{F}(\cdot, A)$; thus, (iv) is satisfied. By the properties of Γ-convergence $F(\cdot, A)$ is lower semicontinuous with respect to the $L^p(\Omega; \mathbb{R}^m)$ topology (see for instance [1, Proposition 1.28]). Thus, Proposition 3.3 yields the weak lower semicontinuity of $F(\cdot, A)$ with respect to the $W^{1,p}(\Omega; \mathbb{R}^m)$ topology. By its definition, $\tilde{F}(\cdot, A)$ inherits the same property, thus (v) holds. As a consequence of Propositions 5.1 and 5.2, $\tilde{F}(u, \cdot)$ is subadditive, superadditive on disjoint sets and inner regular. Hence, by the De Giorgi-Letta measure criterion (see [2]), $\tilde{F}(u, \cdot)$ is the restriction on $\mathcal{A}(\Omega)$ of a Borel measure, thus (ii) is satisfied and the thesis follows.

Step 2   Assume that only (H0') holds. For any $n \in \mathbb{N}$ let $F_{\varepsilon,n} : L^p(\Omega; \mathbb{R}^m) \times \mathcal{A}(\Omega) \to [0, +\infty)$ be defined by

$$F_{\varepsilon,n}(u, A) := F_\varepsilon(u, A) + \frac{1}{n} G_\varepsilon[\psi_{\varepsilon,1}](u, A).$$

Since the family of functionals $F_{\varepsilon,n}$ satisfies (H0)–(H2) for every $n \in \mathbb{N}$, by Step 1 and a diagonalization argument there exist a subsequence (ε_{j_k}) and a non increasing sequence of functions $f_n : \Omega \times \mathbb{R}^{m \times d} \to [0, +\infty)$, $n \in \mathbb{N}$, quasiconvex in the second variable and satisfying (5.1), such that

$$\Gamma(L^p)\text{-} \lim_{k \to \infty} F_{\varepsilon_{j_k}, n}(u, A) = \begin{cases} \displaystyle\int_A f_n(x, \nabla u)\, dx & \text{if } u \in W^{1,p}(A, \mathbb{R}^m), \\ +\infty & \text{otherwise.} \end{cases}$$

for any $A \in \mathcal{A}^{\text{reg}}(\Omega)$. We claim that (5.2) holds with $f_0(x, M) := \inf_{n \in \mathbb{N}} f_n(x, M)$. Indeed, since $F_\varepsilon \le F_{\varepsilon,n}$ for every $\varepsilon > 0$ and $n \in \mathbb{N}$, we have

$$F''(u, A) \le F(u, A) := \begin{cases} \displaystyle\int_A f_0(x, \nabla u)\, dx & \text{if } u \in W^{1,p}(A, \mathbb{R}^m), \\ +\infty & \text{otherwise.} \end{cases}$$

It remains to prove that

$$F(u, A) \le F'(u, A). \tag{5.16}$$

By Proposition 3.3, it suffices to prove (5.16) for $u \in W^{1,p}(A; \mathbb{R}^m)$. Let then $u_k \to u$ in $L^p(\Omega; \mathbb{R}^m)$ and be such that

$$\liminf_{k \to +\infty} F_{\varepsilon_{j_k}}(u_k, A) = F'(u, A).$$

Given $A' \Subset A$, by (H0') we may assume that $G_{\varepsilon_{j_k}}[\psi_{\varepsilon,1}](u_k, A')$ is uniformly bounded. Hence

$$\int_{A'} f_0(x, \nabla u)\, dx \le \int_{A'} f_n(x, \nabla u)\, dx \le \liminf_{k \to +\infty} F_{\varepsilon_{j_k}, n}(u_k, A') \le F'(u, A) + \frac{C}{n}.$$

Thus (5.16) follows from the arbitrariness of $n \in \mathbb{N}$ and $A' \Subset A$.

The proof is then complete. $\qquad\qquad\qquad\qquad\qquad\qquad\qquad\qquad\qquad\quad\ \square$

Remark 5.3 Theorem 5.1 clearly applies also to the truncated energies F_ε^T. Suppose that for any $(u, A) \in W^{1,p}(\Omega; \mathbb{R}^m) \times \mathcal{A}^{\mathrm{reg}}(\Omega)$ and $T > 0$ there exists

$$\Gamma\text{-}\lim_{\varepsilon \to 0} F_\varepsilon^T(u, A) = \int_A f_0^T(x, \nabla u(x))dx. \tag{5.17}$$

Then, by Lemma 5.1 and monotone convergence, we infer that

$$\Gamma\text{-}\lim_{\varepsilon \to 0} F_\varepsilon(u, A) = \int_A f_0(x, \nabla u(x))dx$$

where for almost every $x_0 \in \Omega$ and every $M \in \mathbb{R}^{m \times d}$

$$f_0(x_0, M) := \lim_{T \to +\infty} f_0^T(x_0, M). \tag{5.18}$$

5.5 Convergence of Minimum Problems

In this section we prove the convergence of minimum problems under Dirichlet boundary conditions.

Definition 5.2 (Boundary-Value Problems) For any $g \in L_{\mathrm{loc}}^p(\mathbb{R}^d; \mathbb{R}^m)$, $A \in \mathcal{A}^{\mathrm{reg}}(\Omega)$ and $r > 0$ we set

$$\mathcal{D}^{r,g}(A) := \{u \in L^p(\Omega; \mathbb{R}^m) : u(x) = g(x) \text{ for a.e. } x \in \Omega, \text{ dist}(x, \mathbb{R}^d \backslash A) < r\} \tag{5.19}$$

and define the functionals $F_\varepsilon^{r,g} : L^p(\Omega; \mathbb{R}^d) \times \mathcal{A}^{\mathrm{reg}}(\Omega) \to [0, +\infty)$

$$F_\varepsilon^{r,g}(u, A) := \begin{cases} F_\varepsilon(u, A) & \text{if } u \in \mathcal{D}^{\varepsilon r, g}(A) \\ +\infty & \text{otherwise.} \end{cases} \tag{5.20}$$

When dealing with the affine function $g(x) = Mx$ for some $M \in \mathbb{R}^{m \times d}$ we will use the notation $\mathcal{D}^{r,M}$ and $F_\varepsilon^{r,M}$.

Proposition 5.3 *Let $A \in \mathcal{A}^{\mathrm{reg}}(\Omega)$, let $F_\varepsilon(\cdot, A)$ be defined by (2.3) and assume that (H0)–(H2) hold. Let $\varepsilon_j \to 0$ and let $f_0 : \Omega \times \mathbb{R}^{m \times d} \to [0, +\infty)$ be such that*

$$\Gamma(L^p)\text{-}\lim_{j \to +\infty} F_{\varepsilon_j}(u, A) = \begin{cases} \displaystyle\int_A f_0(x, \nabla u(x)) \, dx & \text{if } u \in W^{1,p}(A; \mathbb{R}^m) \\ +\infty & \text{otherwise} \end{cases} =: F(u, A).$$

Then, given $g \in W^{1,p}_{\mathrm{loc}}(\mathbb{R}^d; \mathbb{R}^m)$ and $r > 0$, the corresponding sequence of functionals $F_{\varepsilon_j}^{r,g}$ defined in (5.20) Γ-converges with respect to the $L^p(\Omega; \mathbb{R}^m)$ topology to the functional

$$F^g(u, A) := \begin{cases} F(u, A) & \text{if } u - g \in W^{1,p}_0(A; \mathbb{R}^m) \\ +\infty & \text{otherwise.} \end{cases} \tag{5.21}$$

Proof Since $F_\varepsilon^{r,g}(u, A) \geq F_\varepsilon(u, A)$, in order to prove the Γ-lim inf inequality it suffices to show that if $u_j \to u$ in $L^p(A; \mathbb{R}^m)$ and $\sup_j F_{\varepsilon_j}^{r,g}(u_j, A)$ is finite, then $u - g \in W^{1,p}_0(A; \mathbb{R}^m)$. Denote by $\tilde{u}_j$ and $\tilde{u}$ the extension of u_j and u on the whole $\mathbb{R}^d$ obtained by setting $\tilde{u}_j = g$, $j \in \mathbb{N}$, and $\tilde{u} = g$ on $\mathbb{R}^d \backslash A$. Let $\tilde{A}$ be an open set such that $\tilde{A} \supseteq A$ and note that, by (2.9), for every $r' \leq \min\{r_0, r/2\}$ we have

$$G_{\varepsilon_j}^{r'}(\tilde{u}_j, \tilde{A}) \leq G_{\varepsilon_j}^{r'}(u_j, A_\varepsilon^{r/2}) + G_{\varepsilon_j}^{r'}(g, \tilde{A} \backslash \overline{A_{\varepsilon_j}^r})$$

$$\leq C(F_{\varepsilon_j}(u_j, A) + |A|) + G_{\varepsilon_j}^{r'}(g, \tilde{A} \backslash \overline{A_{\varepsilon_j}^r}) \leq C,$$

where $A_\varepsilon^\rho := \{x \in A : \mathrm{dist}(x, \Omega \backslash A) > \varepsilon \rho\}$. Since $\tilde{u}_j \to \tilde{u}$ in $L^p(\tilde{A}; \mathbb{R}^m)$, by Proposition 3.1 we get that $\tilde{u} \in W^{1,p}(\tilde{A}; \mathbb{R}^m)$ and thus $u - g \in W^{1,p}_0(A; \mathbb{R}^m)$.

By a density argument it suffices to prove the Γ-lim sup inequality for $u \in W^{1,p}(\Omega; \mathbb{R}^m)$ such that $\mathrm{spt}(u - g) \Subset A$. Given such a u, let u_j converge to u in $L^p(\Omega; \mathbb{R}^m)$ such that

$$\lim_{j \to \infty} F_{\varepsilon_j}(u_j, A) = F(u, A).$$

With an argument analogous to the one used in the proof of Propositions 5.1 and 5.2, given $\delta > 0$, we can find a suitable cut-off function φ_j with $\mathrm{spt}(\varphi_j) \Subset A$ such that, having set $v_j := \varphi_j u_j + (1 - \varphi_j)u$, we have that v_j still converge to u in $L^p(\Omega; \mathbb{R}^m)$ and

$$F_{\varepsilon_j}(v_j, A) \le F_{\varepsilon_j}(u_j, A) + \delta.$$

Since $v_j \in \mathcal{D}^{\varepsilon_j r, g}(A)$ for j large enough, we get

$$\limsup_{j \to \infty} F_{\varepsilon_j}^{r, g}(v_j, A) \le F(u, A) + \delta$$

and the arbitrariness of δ leads to the desired inequality. $\qquad\square$

As a consequence of Propositions 5.3, 4.1 and Theorem 4.2, we derive the following result of convergence of minimum problems with Dirichlet boundary data.

Proposition 5.4 (Convergence of Boundary-Value Problems) *Under the assumptions of Proposition 5.3 there holds*

$$\lim_{j \to \infty} \inf\{F_{\varepsilon_j}(u, A) : u \in \mathcal{D}^{\varepsilon_j r, g}(A)\} = \min\{F(u, A) : u - g \in W_0^{1, p}(A; \mathbb{R}^m)\}.$$

$$(5.22)$$

Moreover, if $u_j \in \mathcal{D}^{\varepsilon_j r, g}(A)$ is a converging sequence such that

$$\lim_{j \to \infty} F_{\varepsilon_j}(u_j, A) = \lim_{j \to \infty} \inf\{F_{\varepsilon_j}(u, A) : u \in \mathcal{D}^{\varepsilon_j r, g}(A)\},$$

then its limit is a minimizer for $\min\{F(u, A) : u - g \in W_0^{1, p}(A; \mathbb{R}^m)\}$.

Proof Note that

$$\inf\{F_{\varepsilon_j}(u, A) : u \in \mathcal{D}^{\varepsilon_j r, g}(A)\} \le F_{\varepsilon_j}(g, A) \le C.$$

Hence, by the properties of Γ-convergence (see for instance [1, Theorem 1.21]), we only need to prove the equi-coerciveness of the family $\{F_{\varepsilon_j}^{r, g}(\cdot, A)\}_{\varepsilon_j}$ in the strong $L^p(A; \mathbb{R}^m)$ topology. Let then $\{u_j\}_j \subset L^p(\Omega; \mathbb{R}^m)$ be such that $F_{\varepsilon_j}^{r, g}(u_j, A) \le C$. Reasoning as in the proof of Proposition 5.3, from assumption (H0) and (2.6) we deduce that $G_{\varepsilon_j}^{r'}(u_j, A) \le C$ for every $r' \le \min\{r_0, r/2\}$. Proposition 4.1 yields that

$$\int_\Omega |u_j(x) - g(x)|^p dx \le C G_{\varepsilon_j}^{r'}(u_j - g, A) \le C\left(G_{\varepsilon_j}^{r'}(u_j, A) + \|\nabla g\|_{L^p(\Omega)}^p\right) \le C.$$

Hence, we may apply Theorem 4.2 and deduce that $\{u_j\}_j$ is precompact in the strong $L^p(\Omega; \mathbb{R}^m)$ topology. $\qquad\square$

Remark 5.4 It is worth noting that infima in (5.22) are indeed minima if the functions f_ε are convex in the last variable, and hence F_ε are weakly lower semicontinuous in $L^p(\Omega; \mathbb{R}^m)$ (see also the next section; we refer e.g. to [5] for lower-semicontiuity issues in the context of peridynamics). Note that if f_ε are not convex in the last variable, then the lower-semicontinuous envelope of F_ε is not a double integral [3, 4].

5.6 Euler-Lagrange Equations

We complete the analysis of convergence of minimum problems of the previous section by studying Euler-Lagrange equations associated to functionals F_ε and F, under suitable regularity assumptions on the densities f_ε.

5.6.1 Regularity of Functionals F_ε

We consider functionals $F_\varepsilon : L^p(\Omega; \mathbb{R}^m) \rightarrow [0, +\infty)$ as defined in (2.1). We preliminary note that F_ε are strongly continuous in the $L^p(\Omega; \mathbb{R}^m)$-topology provided that f_ε is continuous in the last variable and (H1) holds with $\psi_{\varepsilon,2}$, $\rho_{\varepsilon,2} \in L^1(\mathbb{R}^d)$ for every $\varepsilon > 0$.

Indeed, let $u_j \rightarrow u$ in $L^p(\Omega; \mathbb{R}^m)$ as $j \rightarrow +\infty$; then, for almost every $\xi \in \mathbb{R}^d$ there holds

$$\lim_{j \rightarrow +\infty} \int_{\Omega_\varepsilon(\xi)} f_\varepsilon\left(x, \xi, \frac{u_j(x + \varepsilon\xi) - u_j(x)}{\varepsilon}\right) dx$$

$$= \int_{\Omega_\varepsilon(\xi)} f_\varepsilon\left(x, \xi, \frac{u(x + \varepsilon\xi) - u(x)}{\varepsilon}\right) dx$$

and (H1) yields

$$\int_{\Omega_\varepsilon(\xi)} f_\varepsilon\left(x, \xi, \frac{u_j(x + \varepsilon\xi) - u_j(x)}{\varepsilon}\right) dx \leq C\|u_j\|_{L^p(\Omega; \mathbb{R}^m)}^p \psi_{\varepsilon,2}(\xi) + |\Omega|\rho_{\varepsilon,2}(\xi).$$

Hence, the Dominated Convergence Theorem implies the continuity of F_ε.

The next result shows that, assuming C^1-regularity in the last variable of the density and a control of the growth of $\nabla_z f_\varepsilon$, the functionals F_ε are indeed differentiable in the Gateaux sense.

Lemma 5.2 *Let f_ε be C^1 regular in the last variable, and let (H1) hold with $\psi_{\varepsilon,2}$, $\rho_{\varepsilon,2} \in L^1(\mathbb{R}^d)$. If*

$$|\nabla_z f_\varepsilon(x, \xi, z)| \leq \psi(\xi)|z|^{p-1} + \rho(\xi), \tag{5.23}$$

for some $\rho \in L^1(\mathbb{R}^d)$ and $\psi \in L^\infty(\mathbb{R}^d)$, then F_ε is Gateaux differentiable and its Gateaux differential is given by

$$DF_\varepsilon(u)$$
$$= \frac{1}{\varepsilon^{d+1}} \int_\Omega \nabla_z f_\varepsilon\left(y, \frac{x-y}{\varepsilon}, \frac{u(x)-u(y)}{\varepsilon}\right) - \nabla_z f_\varepsilon\left(x, \frac{y-x}{\varepsilon}, \frac{u(y)-u(x)}{\varepsilon}\right) dy.$$
$$(5.24)$$

Proof Let $h_{\varepsilon,u}(x)$ denote the right-hand side of (5.24). For every $v \in L^p(\Omega; \mathbb{R}^m)$, exploiting the symmetric roles of x and y we get

$$\lim_{t \to 0} \frac{F_\varepsilon(u+tv) - F_\varepsilon(u)}{t}$$
$$= \frac{1}{\varepsilon^d} \int_{\Omega \times \Omega} \langle \nabla_z f_\varepsilon\left(x, \frac{y-x}{\varepsilon}, \frac{u(y)-u(x)}{\varepsilon}\right), \frac{v(y)-v(x)}{\varepsilon} \rangle dx\, dy$$
$$= \int_\Omega h_{\varepsilon,u}(x) v(x)\, dx.$$

Condition (5.23) yields that

$$|h_{\varepsilon,u}(x)| \le \frac{2}{\varepsilon^{d+1}} \int_\Omega \psi\left(\frac{y-x}{\varepsilon}\right) \left|\frac{u(y)-u(x)}{\varepsilon}\right|^{p-1} + \rho\left(\frac{y-x}{\varepsilon}\right) dy.$$

If $p \ge 2$ we get

$$|h_{\varepsilon,u}(x)| \le \frac{C}{\varepsilon^{d+p}}\left(|u(x)|^{p-1} + \|u\|_{L^p(\Omega;\mathbb{R}^m)} + \varepsilon^d\right);$$

if instead $1 < p < 2$ then Jensen's inequality implies

$$|h_{\varepsilon,u}(x)| \le \frac{C}{\varepsilon^{d+p}}\left(|u_\Omega - u(x)|^{p-1} + \varepsilon^d\right).$$

In both cases, $h_{\varepsilon,u} \in L^{p-1}(\Omega; \mathbb{R}^m)$; thus, F_ε is Gateaux differentiable and (5.24) holds true. □

We remark that, in the case of f_ε convex in the last variable, (5.23) is a consequence of (H1). We will use this fact in the next section.

5.6.2 Relations with Minimum Problems

Provided suitable regularity (and convexity) conditions on f_ε, minimizers for functionals F_ε are weak solutions to associated Euler-Lagrange equations. In particular, the convergence of minimum problems proved in Sect. 5.5 implies convergence of solutions to the Euler-Lagrange equations of the Γ-limit, in the strictly convex case.

Proposition 5.5 *Let f_ε be convex and of class C^1 in the last variable, and let* (H0) *and* (H1) *hold with $\psi_{\varepsilon,2} \in L^1(\mathbb{R}^d) \cap L^\infty(\mathbb{R}^d)$ and $\rho_{\varepsilon,2} \in L^1(\mathbb{R}^d)$. Let $r > 0$ and $g \in L^p(\Omega; \mathbb{R}^m)$ be given and let $\mathcal{D}^{\varepsilon r, g}(\Omega)$ be as in Definition 5.2. Then there exist solutions to the minimum problem*

$$\min_{u \in \mathcal{D}^{\varepsilon r, g}(\Omega)} F_\varepsilon(u) \qquad (5.25)$$

and every minimizer $u_\varepsilon \in \mathcal{D}^{\varepsilon r, g}(\Omega)$ solves

$$\int_\Omega \nabla_z f_\varepsilon\left(y, \frac{x-y}{\varepsilon}, \frac{u(x)-u(y)}{\varepsilon}\right) - \nabla_z f_\varepsilon\left(x, \frac{y-x}{\varepsilon}, \frac{u(y)-u(x)}{\varepsilon}\right) dy = 0$$

$$(5.26)$$

for almost every $x \in \Omega(\varepsilon r)$, where $\Omega(\varepsilon r) = \{x \in \Omega : \text{dist}(x, \mathbb{R}^d \setminus \Omega) > \varepsilon r\}$.

Proof By the strong continuity and convexity, F_ε is lower-semicontinuous with respect to the weak $L^p(\Omega; \mathbb{R}^m)$ topology. By assumption (H0) and applying Poincaré inequality (Proposition 4.1) as done in the proof of Proposition 5.3, the set

$$\{u \in \mathcal{D}^{\varepsilon r, g}(\Omega) : F_\varepsilon(u) \leq C\}$$

is bounded and therefore F_ε is weakly coercive. This ensures that the minimum problem (5.25) has solutions thanks to the Direct Method of the Calculus of Variations.

The convexity of f_ε and assumption (H1) implies (5.23) with $\psi = \psi_{\varepsilon,2}$; thus, F_ε is Gateaux differentiable and (5.24) holds. The restriction of F_ε to $\mathcal{D}_\varepsilon^{r,g}(\Omega)$ is also differentiable and its differential is defined on $\mathcal{D}^{\varepsilon r, 0}(\Omega)$; thus, by minimality

$$\int_\Omega \langle (DF_\varepsilon(u_\varepsilon))(x), v(x) \rangle dx = 0$$

for every $v \in \mathcal{D}^{\varepsilon r, 0}(\Omega)$ and the result follows. $\square$

We recall well-known properties of local functionals as those that are cluster points (in the sense of Γ-convergence) of F_ε.

Remark 5.5 Consider $F : W^{1,p}(\Omega; \mathbb{R}^m) \to [0, +\infty)$ as defined in the statement of Theorem 5.1; that is,

$$F(u) = \int_\Omega f_0(x, \nabla u(x)) \, dx$$

where $f_0 : \Omega \times \mathbb{R}^{m \times d} \to [0, +\infty)$ is quasiconvex in the second variable and complies with the growth condition (5.1). Then, an application of the Direct Method of the Calculus of Variations in the weak $W^{1,p}(\Omega)$-topology yields the existence of minimizers for the problem

$$\min\{F(u) : u - g \in W_0^{1,p}(\Omega; \mathbb{R}^m)\},$$

where $g \in W^{1,p}(\Omega; \mathbb{R}^m)$ is given.

If we assume in addition that f is C^1 in the last variable and that

$$|\nabla_M f_0(x, M)| \le C(|M|^{p-1} + 1)$$

then F is Gateaux differentiable on $W^{1,p}(\Omega; \mathbb{R}^m)$. Indeed, for every $v \in W_0^{1,p}(\Omega; \mathbb{R}^m)$

$$\lim_{t \to 0} \frac{F(u + tv) - F(u)}{t} = \int_\Omega \langle \nabla_M f_0(x, \nabla u(x)), \nabla v(x) \rangle dx$$

which is linear in v and bounded thanks to the growth condition. Let u_0 be a minimizer of F, then applying the divergence theorem to the right-hand side above we get by minimality that

$$\begin{cases} \mathrm{Div}(\nabla_M f(\cdot, \nabla u_0)) = 0 & \text{in } \Omega \\ u_0 = g & \text{on } \partial\Omega \end{cases} \tag{5.27}$$

in the sense of distribution.

Proposition 5.6 Let f_ε be strictly convex and of class C^1 in the last variable and let (H0)–(H2) hold with $\psi_{\varepsilon,2} \in L^1(\mathbb{R}^d) \cap L^\infty(\mathbb{R}^d)$ and $\rho_{\varepsilon,2} \in L^1(\mathbb{R}^d)$. Let $r > 0$ and $g \in W^{1,p}(\Omega; \mathbb{R}^m)$ be given and let $\mathcal{D}^{\varepsilon r,g}(\Omega)$ be as in Definition 5.2. Then Eqs. (5.26) and (5.27) are uniquely solved by u_ε and u respectively and u_ε converges to u in the weak $W^{1,p}(\Omega; \mathbb{R}^m)$-topology as $\varepsilon \to 0$.

Proof By Theorem 5.1 and the convexity of f_ε we have

$$\Gamma\text{-}\lim_{\varepsilon \to 0} F_\varepsilon(u) = F(u) = \int_\Omega f_0(x, \nabla u(x)) dx$$

for some f_0 strictly convex in the second variable and complying with (5.1). From Proposition 5.5 and Remark 5.5 and the strict convexity of f_ε and f_0 the equations have unique (weak) solutions u_ε and u_0 respectively such that

$$F_\varepsilon(u_\varepsilon) = \min_{\mathcal{D}^{\varepsilon r,g}(\Omega)} F_\varepsilon \quad \text{and} \quad F(u_0) = \min_{g + W_0^{1,p}(\Omega;\mathbb{R}^m)} F.$$

In the end, we remark that the family $F_\varepsilon^{r,g}$ is equi-coercive, as shown in the proof of Proposition 5.4. This, thanks to the convergence of minimum problems provided by Proposition 5.4, implies that for any $\varepsilon_j \to 0$ there exists a subsequence $\{\varepsilon_j'\}$ such that $u_{\varepsilon_j'}$ converges to u_0, and the result is proved. □

References

1. Braides, A.: Γ-Convergence for Beginners. Oxford Lecture Series in Mathematics and Its Applications, vol. 22. Oxford University Press, Oxford (2002)
2. Braides, A., Defranceschi, A.: Homogenization of Multiple Integrals. Oxford Lecture Series in Mathematics and Its Applications, vol. 12. The Clarendon Press/Oxford University Press, New York (1998)
3. Kreisbeck, C., Zappale, E.: Loss of double-integral character during relaxation. SIAM J. Math. Anal. **53**, 351–385 (2021)
4. Mora-Corral, C., Tellini, A.: Relaxation of a scalar nonlocal variational problem with a double-well potential. Calc. Var. Partial Differ. Equ. **59**(67) (2020)
5. Pedregal, P.: Weak lower semicontinuity and relaxation for a class of non-local functionals. Rev. Mat. Complut. **29**, 485–495 (2016)

Chapter 6
Periodic Homogenization

Abstract In this chapter we study the homogenization of convolution functionals under an assumption of periodicity in the space variable of the energy densities. The limit energy density is characterized by an asymptotic nonlocal homogenization formula, which reduces to a non-local cell-problem formula when the energy density is convex in the last variable. In the case of homogeneous integrands the homogenization formula simplify only in the convex case. In the last sections of the chapter we prove the homogenization result in perforated domains using an extension theorem.

Keywords Periodic homogenization · Asymptotic formula · Cell-problem formula · Relaxation · Extension theorems · Lipschitz domains · Perforated domains

6.1 A Homogenization Theorem

Let $f : \mathbb{R}^d \times \mathbb{R}^d \times \mathbb{R}^m \to [0, +\infty)$ be a Borel function such that $f(\cdot, \xi, z)$ is $[0, 1]^d$-periodic in the first variable for every $\xi \in \mathbb{R}^d$ and $z \in \mathbb{R}^m$. Throughout this chapter, we will assume that

$$f_\varepsilon(x, \xi, z) = f\left(\frac{x}{\varepsilon}, \xi, z\right) \tag{6.1}$$

in (2.1). In this setting, assumptions (H0)–(H2) are straightforward consequence of the following growth conditions on f: for a.e. $x \in \mathbb{R}^d$ and every $z \in \mathbb{R}^m$

$$\psi_1(\xi)|z|^p - \rho_1(\xi) \leq f(x, \xi, z), \quad \text{for a.e. } \xi \in B_{r_0}, \tag{6.2}$$

$$f(x, \xi, z) \leq \psi_2(\xi)|z|^p + \rho_2(\xi), \quad \text{for a.e. } \xi \in \mathbb{R}^d, \tag{6.3}$$

© The Author(s), under exclusive license to Springer Nature Singapore Pte Ltd. 2023   59
R. Alicandro et al., *A Variational Theory of Convolution-Type Functionals*,
SpringerBriefs on PDEs and Data Science,
https://doi.org/10.1007/978-981-99-0685-7_6

where $\rho_1, \psi_1 : B_{r_0} \to [0, +\infty)$ and $\psi_2, \rho_2 : \mathbb{R}^d \to [0, +\infty)$ are such that

$$\int_{B_{r_0}} \rho_1(\xi)\, d\xi + \int_{\mathbb{R}^d} \rho_2(\xi)\, d\xi < +\infty, \qquad \int_{\mathbb{R}^d} \psi_2(\xi)|\xi|^p\, d\xi < +\infty, \qquad (6.4)$$

$$\psi_1(\xi) \geq c_0, \qquad \text{for a.e. } \xi \in B_{r_0} \qquad\qquad\qquad (6.5)$$

requesting additionally that

$$\text{if } \int_{B_{r_0}} \psi_2(\xi)\, d\xi = +\infty \text{ then there exists } c_1 > 0 \text{ such that}$$
$$\qquad\qquad\qquad\qquad\qquad\qquad\qquad\qquad\qquad\qquad (6.6)$$
$$\psi_2(\xi) \leq c_1\psi_1(\xi) \text{ for a.e. } \xi \in B_{r_0}.$$

In the sequel we will use the notation $Q_R(x_0) = x_0 + (0, R)^d$ and the shorthand Q_R if $x_0 = 0$ The main result of this chapter is stated in the following theorem.

Theorem 6.1 (Homogenization Theorem) *Let F_ε be defined by (2.1), with f_ε given by (6.1), and let (6.2)–(6.6) be satisfied. Then for every $M \in \mathbb{R}^{m \times d}$ the limit (asymptotic homogenization formula)*

$$f_{\text{hom}}(M) := \lim_{R \to \infty} \frac{1}{R^d} \inf \Big\{ \int_{Q_R} \int_{Q_R} f(x, y - x, v(y) - v(x))dx\, dy$$

$$: v \in \mathcal{D}^{1,M}(Q_R) \Big\}, \qquad\qquad\qquad (6.7)$$

where $\mathcal{D}^{1,M}(Q_R)$ is defined by (5.19), exists and defines a quasiconvex function $f_{\text{hom}} : \mathbb{R}^{m \times d} \to [0, +\infty)$ satisfying

$$c(|M|^p - 1) \leq f_{\text{hom}}(M) \leq C(|M|^p + 1). \qquad\qquad (6.8)$$

Moreover,

$$\Gamma(L^p)\text{-} \lim_{\varepsilon \to 0} F_\varepsilon(u) = \begin{cases} \int_\Omega f_{\text{hom}}(\nabla u(x))dx & \text{if } u \in W^{1,p}(\Omega; \mathbb{R}^m) \\ +\infty & \text{otherwise.} \end{cases}$$

The proof of Theorem 6.1 relies on the results stated in the following two propositions. The first one provides the independence of the limit energy densities on the space variable, the second one the existence of the limit in (6.7) in the case of truncated energies.

Proposition 6.1 *Under the assumptions of Theorem 6.1, let $\varepsilon_j \to 0$ and let $f_0 :$ $\Omega \times \mathbb{R}^{m \times d} \to [0, +\infty)$ be a Carathéodory function such that for every $A \in \mathcal{A}^{\mathrm{reg}}(\Omega)$ and $u \in W^{1,p}(A; \mathbb{R}^m)$ there holds*

$$\Gamma(L^p)\text{-}\lim_{j \to +\infty} F_{\varepsilon_j}(u, A) = \int_A f_0(x, \nabla u(x)) dx.$$

Then f_0 is independent of the first variable.

Proof It is sufficient to prove that

$$F(Mx, B_r(y)) = F(Mx, B_r(y')) \tag{6.9}$$

for every $M \in \mathbb{R}^{m \times d}$, $y, y' \in \Omega$ and $r > 0$ such that $B_r(y), B_r(y') \subset \Omega$. We will prove that $F(Mx, B_{r'}(y)) \le F(Mx, B_r(y'))$ for all $r' < r$. By the inner regularity of $F(Mx, \cdot)$ provided by Proposition 5.2 we get (6.9) by switching the roles of y and y'.

Let $u_j \to Mx$ in $L^p(B_r(y'); \mathbb{R}^m)$ be such that

$$\lim_{j \to +\infty} F_{\varepsilon_j}(u_j, B_r(y')) = F(Mx, B_r(y')),$$

and set $v_j(x) := u_j\left(x - \varepsilon_j \left\lfloor \dfrac{y - y'}{\varepsilon_j} \right\rfloor\right) + \varepsilon_j M \left\lfloor \dfrac{y - y'}{\varepsilon_j} \right\rfloor$. Since $B_{r'}(y) - \varepsilon_j \left\lfloor \dfrac{y - y'}{\varepsilon_j} \right\rfloor \subset B_r(y')$ when ε_j is small enough, $v_j \to Mx$ in $L^p(B_{r'}(y); \mathbb{R}^m)$. Moreover, by the periodicity assumption on f, we have that

$$F_{\varepsilon_j}(v_j, B_{r'}(y))$$

$$= \int_{\mathbb{R}^d} \int_{(B_{r'}(y))_{\varepsilon_j}(\xi)} f\left(\frac{x}{\varepsilon_j}, \xi, \frac{v_j(x + \varepsilon_j \xi) - v_j(x)}{\varepsilon_j}\right) dx \, d\xi$$

$$= \int_{\mathbb{R}^d} \int_{(B_{r'}(y))_{\varepsilon_j}(\xi)} f\left(\frac{x}{\varepsilon_j} - \left\lfloor \frac{y - y'}{\varepsilon_j} \right\rfloor, \xi, \frac{v_j(x + \varepsilon_j \xi) - v_j(x)}{\varepsilon_j}\right) dx \, d\xi.$$

and, through the change of variable $x = x' + \varepsilon_j \lfloor (y - y')/\varepsilon_j \rfloor$, we get

$$F_{\varepsilon_j}(v_j, B_{r'}(y)) \le \int_{\mathbb{R}^d} \int_{(B_r(y'))_{\varepsilon_j}(\xi)} f\left(\frac{x'}{\varepsilon_j}, \xi, \frac{u_j(x' + \varepsilon_j \xi) - u_j(x')}{\varepsilon_j}\right) dx' \, d\xi$$

$$= F_{\varepsilon_j}(u_j, B_r(y')).$$

Finally, letting $j \to +\infty$, we obtain

$$F(Mx, B_{r'}(y)) \leq \liminf_{j \to +\infty} F_{\varepsilon_j}(v_j, B_{r'}(y))$$

$$\leq \lim_{j \to +\infty} F_{\varepsilon_j}(u_j, B_r(y')) = F(Mx, B_r(y'))$$

and the claim.   □

Proposition 6.2 *Let $f : \mathbb{R}^d \times \mathbb{R}^d \times \mathbb{R}^m \to [0, +\infty)$ be a Borel function $[0, 1]^d$-periodic in the first variable such that assumptions (6.3) and (6.4) hold. Let $T > 0$ and set*

$$f^T(x, \xi, z) = \begin{cases} f(x, \xi, z) & \text{if } |\xi| < T, \\ 0 & \text{otherwise.} \end{cases} \tag{6.10}$$

Then, for every $r \geq T$ the limit

$$f_{\text{hom}}^T(M) := \lim_{R \to \infty} \frac{1}{R^d} \inf \left\{ \int_{Q_R} \int_{Q_R} f^T(x, y - x, v(y) - v(x)) dx \, dy \right.$$

$$\left. : v \in \mathcal{D}^{r,M}(Q_R) \right\}, \tag{6.11}$$

exists and it is finite for every $M \in \mathbb{R}^{m \times d}$, with $\mathcal{D}^{r,M}(Q_R)$ defined as in Definition 5.2.

Proof For every $R > 0$ we set

$$F_1^T(v, Q_R) := \int_{Q_R} \int_{Q_R} f^T(x, y - x, v(y) - v(x)) dx \, dy,$$

$$H_R(M) := \frac{1}{R^d} \inf \left\{ F_1^T(v, Q_R) : v \in \mathcal{D}^{r,M}(Q_R) \right\}$$

and let $u_R \in \mathcal{D}^{r,M}(Q_R)$ be such that

$$\frac{1}{R^d} F_1^T(u_R, Q_R) \leq H_R(M) + \frac{1}{R}.$$

For any $S > R$, define $S_R := \lfloor S/R \rfloor R$ and

$$\mathcal{L}_{R,S} := R \left\{ 0, 1, \ldots, \left\lfloor \frac{S}{R} \right\rfloor - 1 \right\}^d, \quad \mathcal{S}_{R,S} := \bigcup_{h \in \mathcal{L}_{R,S}} \partial(h + Q_R).$$

Hence, we define

$$v_S(x) := \begin{cases} u_R(x - h) + Mh & \text{if } x \in h + Q_R, h \in \mathcal{L}_{R,S} \\ Mx & \text{otherwise.} \end{cases}$$

Note that $f^T(x, y - x, v_S(y) - v_S(x)) \neq 0$ only if $x, y \in h + Q_R$, for some $h \in \mathcal{L}_{R,S}$, or $x, y \in S_{R,S}^T$, where

$$S_{R,S}^T := \{x \in Q_S : \text{dist}(x, S_{R,S}) < T\} \cup (Q_S \setminus Q_{S_R}).$$

In the latter case, since $r \geq T$, $v_S(x) = Mx$ and $v_S(y) = My$. Hence, from the definition of v_S we get

$$F_1^T(v_S, Q_S) \leq \int_{S_{R,S}^T} \int_{S_{R,S}^T} f^T(x, y - x, My - Mx) dx\, dy$$

$$+ \sum_{h \in \mathcal{L}_{R,S}} \int_{h+Q_R} \int_{h+Q_R} f^T(x, y - x, u_R(y - h) - u_R(x - h)) dx\, dy.$$

$$(6.12)$$

By the periodicity of $f(\cdot, \xi, z)$ we have that

$$\sum_{k \in \mathcal{L}_{R,S}} \int_{h+Q_R} \int_{h+Q_R} f^T(x, y - x, u_R(y - h) - u_R(x - h)) dx\, dy$$

$$= \left\lfloor \frac{S}{R} \right\rfloor^d \left(\int_{Q_R} \int_{Q_R} f^T(x, y - x, u_R(y) - u_R(x)) dx\, dy \right).$$

$$(6.13)$$

By (6.3) and (6.4), we get

$$\int_{S_{R,S}^T} \int_{S_{R,S}^T} f^T(x, y - x, My - Mx) dx\, dy \leq C(|M|^p + 1)(T + 1)S^{d-1}\left(\frac{S}{R} + R\right).$$

$$(6.14)$$

Gathering (6.12)–(6.14), from the definition of u_R we obtain

$$F_1^T(v_S, Q_S) \leq \left\lfloor \frac{S}{R} \right\rfloor^d \left(R^d H_R(M) + R^{d-1} \right) + C(|M|^p + 1)(T + 1)S^{d-1}\left(\frac{S}{R} + R\right).$$

$$(6.15)$$

Finally, by using v_S as a test function in the definition of $H_S(M)$, (6.15) gives

$$H_S(M) \leq \frac{R^d}{S^d} \left\lfloor \frac{S}{R} \right\rfloor^d H_R(M) + \frac{R^{d-1}}{S^d} \left\lfloor \frac{S}{R} \right\rfloor^d + C(|M|^p + 1)(T + 1)\left(\frac{1}{R} + \frac{R}{S}\right).$$

By taking the limit first as $S \to +\infty$ and then as $R \to +\infty$ we get

$$\limsup_{S \to +\infty} H_S(M) \leq \liminf_{R \to +\infty} H_R(M)$$

that yields the existence of the limit. Finally, $H_R(M) \leq F_1^T(Mx, Q_R)/R^d \leq C(|M|^p + 1)$, hence the limit is finite. $\qquad\square$

Proof (of Theorem 6.1) We split the proof in two steps, dealing first with the case of truncated energies and then with the general case.

Step 1 For $T > 0$ let $F_\varepsilon^T(\cdot, \cdot)$ be as in Definition 5.1. Given $\varepsilon_j \to 0$, by Theorem 5.1 and Proposition 6.1 there exist a subsequence (not relabelled) and $f_0 : \mathbb{R}^{m \times d} \to [0, +\infty)$ such that

$$\Gamma(L^p)\text{-} \lim_{j \to +\infty} F_{\varepsilon_j}^T(u, A) = \int_A f_0(\nabla u(x))\, dx =: F^T(u, A)$$

for every $A \in \mathscr{A}^{\mathrm{reg}}(\Omega)$ and $u \in W^{1,p}(A; \mathbb{R}^m)$. We now prove that $f_0 = f_{\mathrm{hom}}^T$, where f_{hom}^T is defined in (6.11). Since f_0 is quasiconvex and satisfies the growth conditions (5.1), given $x_0 \in \Omega$ and $r > 0$ such that $Q_r(x_0) \subset \Omega$, we have

$$f_0(M) = \frac{1}{r^d} \min\left\{ \int_{Q_r(x_0)} f_0(\nabla u(x))dx : u - Mx \in W_0^{1,p}(Q_r(x_0); \mathbb{R}^m) \right\}$$

for every $M \in \mathbb{R}^{m \times d}$ and then, by Proposition 5.4, we get

$$f_0(M) = \lim_{j \to \infty} \frac{1}{r^d} \inf\left\{ F_{\varepsilon_j}^T(u, Q_r(x_0)) : u \in \mathcal{D}^{s\varepsilon_j, M}(Q_r(x_0)) \right\}$$

for any $s > 0$. Without loss of generality we may assume $x_0 = 0$ and use the notation $Q_r = Q_r(0)$. Setting $v(x) = u(\varepsilon_j x)/\varepsilon_j$ and using the changes of variable $x' = x/\varepsilon_j$ and $y = x' + \xi$, we rescale $F_{\varepsilon_j}^T$ as

$$F_{\varepsilon_j}^T(u, Q_r) = \int_{B_T} \int_{(Q_r)_{\varepsilon_j}(\xi)} f\left(\frac{x}{\varepsilon_j}, \xi, \frac{u(x + \varepsilon_j \xi) - u(x)}{\varepsilon_j}\right) dx\, d\xi$$

$$= \int_{B_T} \int_{(Q_r)_{\varepsilon_j}(\xi)} f\left(\frac{x}{\varepsilon_j}, \xi, v\left(\frac{x}{\varepsilon_j} + \xi\right) - v\left(\frac{x}{\varepsilon_j}\right)\right) dx\, d\xi$$

$$= \varepsilon_j^d \int_{Q_{R_j}} \int_{Q_{R_j}} f^T(x', y - x', v(y) - v(x'))dx'dy,$$

where $R_j := r/\varepsilon_j$ and f^T is defined in (6.10). Thus,

$$f_0(M) = \lim_{j \to \infty} \frac{1}{R_j^d} \inf \left\{ \int_{Q_{R_j}} \int_{Q_{R_j}} f^T(x, y - x, v(y)) \right.$$

$$\left. - v(x))dx\, dy : v \in \mathcal{D}^{s,M}(Q_{R_j}) \right\}.$$

By the arbitrariness of $s > 0$ and Proposition 6.2 we eventually get

$$f_0(M) = f_{\text{hom}}^T(M)$$

$$= \lim_{R \to \infty} \frac{1}{R^d} \inf \left\{ \int_{Q_R} \int_{Q_R} f^T(x, y - x, v(y) - v(x))dx\, dy \right.$$

$$: v \in \mathcal{D}^{1,M}(Q_R) \right\},$$

which in particular proves the claim of Theorem 6.1 when $f \equiv f^T$.

Step 2 By the previous step, Lemma 5.1 and Remark 5.3 we infer that

$$\Gamma(L^p)\text{-}\lim_{\varepsilon \to 0} F_\varepsilon(u, A) = \lim_{T \to +\infty} \Gamma(L^p)\text{-}\lim_{\varepsilon \to 0} F_\varepsilon^T(u, A) = \int_A f_{\text{hom}}^\infty(\nabla u(x))dx$$

for every $A \in \mathcal{A}^{\text{reg}}(\Omega)$ and $u \in W^{1,p}(A; \mathbb{R}^m)$, where for every $M \in \mathbb{R}^{m \times d}$

$$f_{\text{hom}}^\infty(M) := \lim_{T \to +\infty} f_{\text{hom}}^T(M).$$

Hence, the result will follow if we prove that $f_{\text{hom}}^\infty(M) = f_{\text{hom}}(M)$. Set

$$f_{\text{hom}}'(M) = \liminf_{R \to \infty} \frac{1}{R^d} \inf \left\{ \int_{Q_R} \int_{Q_R} f(x, y - x, v(y) - v(x))dx\, dy \right.$$

$$: v \in \mathcal{D}^{1,M}(Q_R) \right\},$$

$$f_{\text{hom}}''(M) = \limsup_{R \to \infty} \frac{1}{R^d} \inf \left\{ \int_{Q_R} \int_{Q_R} f(x, y - x, v(y) - v(x))dx\, dy \right.$$

$$: v \in \mathcal{D}^{1,M}(Q_R) \right\}.$$

Since $f_{\text{hom}}^T(M) \leq f_{\text{hom}}'(M)$ for every $T > 0$, it suffices to prove that $f_{\text{hom}}''(M) \leq f_{\text{hom}}^\infty(M)$. Now, define

$$H_R^T(M) = \frac{1}{R^d} \inf \left\{ \int_{Q_R} \int_{Q_R} f^T(x, y-x, v(y) - v(x)) dx\, dy : v \in \mathcal{D}^{1,M}(Q_R) \right\},$$

$$H_R(M) = \frac{1}{R^d} \inf \left\{ \int_{Q_R} \int_{Q_R} f(x, y-x, v(y) - v(x)) dx\, dy : v \in \mathcal{D}^{1,M}(Q_R) \right\},$$

and let $u_R \in \mathcal{D}^{1,M}(Q_R)$ be such that

$$\frac{1}{R^d} \int_{Q_R} \int_{Q_R} f^T(x, y-x, u_R(y) - u_R(x)) dx\, dy \leq H_R^T(M) + \frac{1}{T}.$$

Note that, by (6.2)–(6.5), we get that

$$\frac{1}{R^d} G_1^{r_0}(u_R, Q_R) \leq C\left(H_R^T(M) + \frac{1}{T} + 1 \right) \leq C\left(\frac{1}{R^d} F_1^T(Mx, Q_R) + \frac{1}{T} + 1 \right)$$
$$\leq C(|M|^p + 1),$$

where C is a constant independent of T and R, so that, by Lemma 4.1, we get that

$$\frac{1}{R^d} \int_{(Q_R)_1(\xi)} |u_R(x + \xi) - u_R(x)|^p dx \leq C(|\xi|^p + 1)(|M|^p + 1).$$

By taking u_R as a test function for the minimum problem defining $H_R(M)$, we then have

$$H_R(M) \leq H_R^T(M) + \frac{1}{T} + \frac{1}{R^d} \int_{B_T^c} \int_{(Q_R)_1(\xi)} f(x, \xi, u_R(x+\xi) - u_R(x)) dx\, d\xi$$

$$\leq H_R^T(M) + \frac{1}{T}$$

$$+ \frac{1}{R^d} \int_{B_T^c} \int_{(Q_R)_1(\xi)} \left(\psi_2(\xi)(|u_R(x+\xi) - u_R(x)|^p) + \rho_2(\xi) \right) dx\, d\xi$$

$$\leq H_R^T(M) + \frac{1}{T} + C(|M|^p + 1) \int_{B_T^c} \psi_2(\xi)(|\xi|^p + 1)\, d\xi + C \int_{B_T^c} \rho_2(\xi)\, d\xi.$$

By assumption (6.3) taking the limit first as $R \to +\infty$ and then as $T \to +\infty$ we get the conclusion.

$\square$

As a straightforward consequence of Theorem 6.1, Propositions 5.3 and 5.4, we deduce the following results about Γ-convergence and convergence of minimum problems for periodic functionals subject to Dirichlet boundary conditions.

Proposition 6.3 *Under the assumptions of Theorem 6.1, given any* $g \in W_{\text{loc}}^{1,p}(\mathbb{R}^d; \mathbb{R}^m)$ *and* $r > 0$, *let* $\{F_\varepsilon^{r,g}(\cdot, \Omega)\}$ *be the family of functionals defined in* (5.20). *Then*

$$\Gamma\text{-}\lim_{\varepsilon \to 0} F_\varepsilon^{r,g}(u, \Omega) = \begin{cases} \displaystyle\int_\Omega f_{\text{hom}}(\nabla u(x))dx & \text{if } u - g \in W_0^{1,p}(\Omega; \mathbb{R}^m) \\ +\infty & \text{otherwise.} \end{cases} \tag{6.16}$$

Proposition 6.4 *Under the assumptions of Theorem 6.1, for any* $g \in W_{\text{loc}}^{1,p}(\mathbb{R}^d; \mathbb{R}^m)$ *and* $r > 0$ *there holds*

$$\lim_{\varepsilon \to 0} \inf\{F_\varepsilon(u) : u \in \mathcal{D}^{r\varepsilon,g}(\Omega)\} = \min\left\{ \int_\Omega f_{\text{hom}}(\nabla u(x))dx \right.$$

$$\left. : u - g \in W_0^{1,p}(\Omega; \mathbb{R}^m) \right\}. \tag{6.17}$$

Moreover, if $\varepsilon_j \to 0$ *and* $u_j \in \mathcal{D}^{r\varepsilon_j,g}(\Omega)$ *is a converging sequence such that*

$$\lim_{j \to \infty} F_{\varepsilon_j}(u_j) = \lim_{j \to \infty} \inf\{F_{\varepsilon_j}(u) : u \in \mathcal{D}^{r\varepsilon_j,g}(\Omega)\},$$

then its limit is a minimizer for $\min\left\{ \int_\Omega f_{\text{hom}}(\nabla u(x))dx : u - g \in W_0^{1,p}(\Omega; \mathbb{R}^m) \right\}$.

6.2 The Convex Case

In this section we show that, analogously to the homogenization of integral functionals, in the convex case the asymptotic formula (6.7) reduces to a cell-problem formula.

Theorem 6.2 (Convex-Homogenization Theorem) *Under the hypotheses of Theorem 6.1, assume in addition that* $f(y, \xi, \cdot)$ *is convex for every* $y, \xi \in \mathbb{R}^d$. *Then the function* f_{hom} *defined by* (6.7) *satisfies for every* $M \in \mathbb{R}^{m \times d}$

$$f_{\text{hom}}(M) = \inf\left\{ \int_{\mathbb{R}^d} \int_{Q_1} f(x, y - x, v(y) - v(x))dx\, dy : v \in \mathcal{D}^{\#,M}(Q_1) \right\}, \tag{6.18}$$

where

$$\mathcal{D}^{\#,M}(Q_1) = \{u \in L_{\text{loc}}^p(\mathbb{R}^d; \mathbb{R}^m) : u - Mx \text{ is } Q_1\text{-periodic}\}.$$

Proof For brevity of notation, we denote by $f^{\#}(M)$ the right-hand side of (6.18). We first prove that $f_{\text{hom}}(M) \leq f^{\#}(M)$. Given $\delta > 0$, let $v \in \mathcal{D}^{\#,M}(Q_1)$ be such that

$$\int_{\mathbb{R}^d} \int_{Q_1} f(x, y - x, v(y) - v(x)) dx\, dy \leq f^{\#}(M) + \delta,$$

and set $u_{\varepsilon}(x) := \varepsilon v(\frac{x}{\varepsilon})$. Then $u_{\varepsilon} \to Mx$ in $L^p(\Omega; \mathbb{R}^m)$ and, by Theorem 6.1 and the periodicity of f, we have

$$|\Omega| f_{\text{hom}}(M) \leq \limsup_{\varepsilon \to 0} F_{\varepsilon}(u_{\varepsilon}) \leq |\Omega|(f^{\#}(M) + \delta).$$

The conclusion follows by the arbitrariness of $\delta > 0$.

It remains to prove that

$$f_{\text{hom}}(M) \geq f^{\#}(M). \tag{6.19}$$

Note that, reasoning as in Step 2 of the proof of Theorem 6.1, we get that

$$\lim_{T \to +\infty} f^{\#,T}(M) = f^{\#}(M),$$

where $f^{\#,T}(M)$ is defined by the right-hand side of (6.18) with f^T in place of f. Hence it suffices to prove (6.19) with f^T. Let $R \in \mathbb{N}$ and, for any function $v \in \mathcal{D}^{T,M}(Q_R)$, let $u \in \mathcal{D}^{\#,M}(Q_1)$ be the function defined by

$$u(x) := \frac{1}{R^d} \sum_{i \in [0,R)^d \cap \mathbb{Z}^d} \tilde{v}(x + i),$$

where $\tilde{v} = \tilde{w}(x) + Mx$ and $\tilde{w}$ denotes the periodic extension of $v(x) - Mx$ outside Q_R. From the convexity of $f(x, \xi, \cdot)$ we get

$$f^{\#,T}(M) \leq \int_{B_T} \int_{Q_1} f(x, \xi, u(x + \xi) - u(x)) dx\, d\xi$$

$$\leq \frac{1}{R^d} \sum_{i \in Q_R \cap \mathbb{Z}^d} \int_{B_T} \int_{i+Q_1} f(x, \xi, \tilde{v}(x + \xi) - \tilde{v}(x)) dx\, d\xi \tag{6.20}$$

$$= \frac{1}{R^d} \int_{B_T} \int_{Q_R} f(x, \xi, \tilde{v}(x + \xi) - \tilde{v}(x)) dx\, d\xi.$$

Since for every $v \in \mathcal{D}^{T,M}(Q_R)$ there holds

$$\int_{B_T} \int_{Q_R \setminus (Q_R)_1(\xi)} f(x, \xi, v(x+\xi) - v(x)) dx \, d\xi$$

$$\leq \int_{B_T} \int_{Q_R \setminus (Q_R)_1(\xi)} f(x, \xi, M\xi) dx \, d\xi \leq C(|M|^p + 1) T R^{d-1},$$

by taking the infimum in (6.20) we get

$$f^{\#,T}(M) \leq \inf \left\{ \frac{1}{R^d} \int_{B_T} \int_{(Q_R)_1(\xi)} f(x, \xi, v(x+\xi) - v(x)) dx \, d\xi \right.$$

$$\left. : v \in \mathcal{D}^{T,M}(Q_R) \right\} + \frac{CT}{R}.$$

Then, passing to the limit as $R \to +\infty$ we obtain the desired inequality. □

Example 6.1 (Quadratic Forms) A well-known property of Γ-convergence is the fact that the Γ-limit of non-negative quadratic forms is still a non-negative quadratic form (see [7, Theorem 11.10]). Hence, under the hypotheses of Theorem 6.1, if $f(x, \xi, z)$ is a non-negative quadratic form of the type

$$f(y, \xi, z) = \langle A(y, \xi) z, z \rangle$$

where $A : \mathbb{R}^d \times \mathbb{R}^d \to \mathbb{R}^{m \times m}$ is $[0, 1]^d$-periodic then

$$f_{\text{hom}}(M) = \langle A_{\text{hom}} M, M \rangle$$

$$= \inf_{v \in \mathcal{D}^{\#,M}(Q_1)} \int_{\mathbb{R}^d} \int_{Q_1} \langle A(x, \xi)(v(x+\xi) - v(x)), v(x+\xi) - v(x) \rangle dx \, d\xi,$$

with $A_{\text{hom}} \in T_2(\mathbb{R}^{m \times d})$, where $T_2(\mathbb{R}^{m \times d})$ denotes the space of $(0, 2)$-tensors on $\mathbb{R}^{m \times d}$.

6.3 Relaxation of Convolution-Type Energies

In the case of energies independent of ε; that is when

$$f_\varepsilon(x, \xi, z) = f(\xi, z), \tag{6.21}$$

energies of the form

$$F_\varepsilon(u) := \int_{\mathbb{R}^d} \int_{\Omega_\varepsilon(\xi)} f\left(\xi, \frac{u(x + \varepsilon\xi) - u(x)}{\varepsilon}\right) dx \, d\xi \tag{6.22}$$

can be regarded as a convolution version of homogeneous functionals and the related homogenization formula

$$f_{\text{hom}}(M) := \lim_{R \to \infty} \frac{1}{R^d} \inf \left\{ \int_{Q_R} \int_{Q_R} f(y - x, v(y) - v(x)) dx \, dy \right.$$

$$\left. : v \in \mathcal{D}^{1,M}(Q_R) \right\} \tag{6.23}$$

can be considered as a kind of *relaxation formula*.

Remark 6.1 If the function f is also convex, then the computation of f_{hom} is trivial, and

$$f_{\text{hom}}(M) = \int_{\mathbb{R}^d} f(\xi, M\xi) \, d\xi. \tag{6.24}$$

Indeed, we can apply Theorem 6.2. After the change of variable $y = x + \xi$, (6.18) reads

$$f_{\text{hom}}(M) = \inf \left\{ \int_{\mathbb{R}^d} \int_{Q_1} f(\xi, v(x + \xi) - v(x)) \, dx \, d\xi : v \in \mathcal{D}^{\#,M}(Q_1) \right\}.$$

Hence, for every $v \in \mathcal{D}^{\#,M}(Q_1)$, Jensen's inequality yields that

$$\int_{\mathbb{R}^d} \int_{Q_1} f(\xi, v(x + \xi) - v(x)) \, dx \, d\xi \geq \int_{\mathbb{R}^d} f(\xi, M\xi) \, d\xi.$$

Thus, by taking the infimum over v we obtain

$$f_{\text{hom}}(M) \geq \int_{\mathbb{R}^d} f(\xi, M\xi) \, d\xi.$$

The converse inequality comes by taking $v(x) = Mx$ as a test function and the claim follows.

A particular case is when $f(\xi, z)$ is a non-negative quadratic form of the type

$$f(\xi, z) = \langle A(\xi)z, z \rangle,$$

in which case

$$f_{\text{hom}}(M) = \int_{\mathbb{R}^d} \langle A(\xi)M\xi, M\xi \rangle \, d\xi.$$

In particular, if $A(\xi) = a(\xi)I$ we recover the result in Theorem 3.1 with $a_\varepsilon(\xi) = a(\xi)$.

More generally, if we do not assume the convexity of $f(\xi, \cdot)$, we have the following bounds

$$\overline{f^{**}}(M) \leq f_{\text{hom}}(M) \leq Q\overline{f}(M), \tag{6.25}$$

where, given $g : \mathbb{R}^d \times \mathbb{R}^m \to [0, +\infty)$, we set

$$\overline{g}(M) := \int_{\mathbb{R}^d} g(\xi, M\xi) \, d\xi \quad M \in \mathbb{R}^{m \times d},$$

$f^{**}(\xi, \cdot)$ denotes the convex envelope of $f(\xi, \cdot)$ and $Q\overline{f}(\cdot)$ denotes the quasi-convex envelope of $\overline{f}(\cdot)$. Indeed the first inequality follows from (6.24), since

$$f^{**}(\xi, z) \leq f(\xi, z), \quad z \in \mathbb{R}^d.$$

To prove the second inequality, notice that, by taking $v(x) = Mx$ as a test function in (6.7), we get

$$f_{\text{hom}}(M) \leq \overline{f}(M),$$

from which the inequality follows, since $f_{\text{hom}}(\cdot)$ is quasi-convex.

In the next two one-dimensional examples we show that both inequalities in (6.25) may be strict.

Example 6.2 ($\overline{f^{**}}(M) < f_{\text{hom}}(M)$) Let F_ε be defined by (2.1), with $d = m = 1$, $\Omega = (0, 1)$ and $f_\varepsilon(x, \xi, z) = f(\xi, z)$, where

$$f(\xi, z) := \begin{cases} f_0\left(\dfrac{z}{\xi}\right) & \text{if } 0 < |\xi| < \frac{1}{2} \\ f_1\left(\dfrac{z}{\xi}\right) & \text{if } \frac{1}{2} \leq |\xi| < 1 \\ f_2\left(\dfrac{z}{\xi}\right) & \text{if } 1 \leq |\xi| < 2 \\ 0 & \text{otherwise,} \end{cases}$$

with

$$f_0(z) = z^2, \quad f_1(z) = (|z| - 1)^2, \quad f_2(z) = \frac{1}{2}z^2.$$

We get that

$$\overline{f^{**}}(z) = z^2 + \begin{cases} z^2 & \text{if } |z| < 1 \\ (|z| - 1)^2 + z^2 & \text{if } |z| \geq 1. \end{cases}$$

We now show that

$$f_{\text{hom}}(z) > \overline{f^{**}}(z) \quad \text{for all } z \in (-1, 0) \cup (0, 1).$$

To this end, we suitably bound F_ε from below with discrete energies whose limit energy density is explicit. More precisely, let $F_\varepsilon^{\text{disc}}(u, (a, b))$ be the functional defined on discrete functions $u : \varepsilon \mathbb{Z} \cap (0, 1) \to \mathbb{R}$ and localized on each interval $(a, b) \subseteq (0, 1)$ as

$$F_\varepsilon^{\text{disc}}(u, (a, b)) = \sum_{j=1}^{2} \sum_{k, k+j\varepsilon \in \varepsilon\mathbb{Z} \cap (a, b)} \varepsilon j f_j \left(\frac{u(k + j\varepsilon) - u(k)}{j\varepsilon} \right)$$

The discrete functions u are identified with their piecewise interpolation on the elementary cells of the lattice $\varepsilon\mathbb{Z}$; that is, with a little abuse of notation,

$$u(x) = u(k) \text{ if } x \in k + (0, \varepsilon), \ k \in \varepsilon\mathbb{Z} \cap (0, 1).$$

Hence $F_\varepsilon^{\text{disc}}$ can be regarded as defined on a subset of $L^2(0, 1)$. In [5] it was proved that

$$\Gamma\text{- } \lim_{\varepsilon \to 0} F_\varepsilon^{\text{disc}}(u, (a, b)) = \int_a^b (g_0)^{**}(u') \, dx, \quad u \in W^{1,2}((a, b)), \tag{6.26}$$

where the Γ-limit is performed with respect to the L^2-convergence and

$$g_0(z) := \frac{1}{2} \inf\{ f_1(z_1) + f_1(z_2) : z_1 + z_2 = z \} + 2 f_2(z).$$

With our choice of f_1 and f_2, $(g_0)^{**}$ can be explicitly computed, giving

$$(g_0)^{**}(z) = \begin{cases} 2z^2 & \text{if } |z| \leq \frac{1}{4} \\ |z| - \frac{1}{8} & \text{if } \frac{1}{4} < |z| < \frac{3}{4} \\ (|z| - 1)^2 + z^2 & \text{if } |z| \geq \frac{3}{4}. \end{cases}$$

Notice that for $z \in (-1, 0) \cup (0, 1)$

$$(g_0)^{**}(z) > z^2.$$

Given $u \in L_{\text{loc}}^1(\mathbb{R})$, set

$$u^{\varepsilon, t}(x) := u(\varepsilon t + \varepsilon \lfloor x/\varepsilon \rfloor).$$

As a consequence of a more general result we will exploit in the following (see (6.29)), we have that if $u_\varepsilon \to u$ in $L^1_{loc}(\mathbb{R})$, then

$$(u_\varepsilon)^{\varepsilon,t} \to u \text{ in } L^1_{loc}(\mathbb{R}) \text{ for a.e. } t \in (0, 1). \tag{6.27}$$

We can now prove the claim. Let $z \in \mathbb{R}$ and let $u_\varepsilon \to zx$ in $L^2(0, 1)$ such that

$$\lim_{\varepsilon \to 0} F_\varepsilon(u_\varepsilon) = f_{\text{hom}}(z).$$

Since $f(-\xi, -z) = f(\xi, z)$, we have

$$F_\varepsilon(u_\varepsilon) \geq 2 \int_0^{\frac{1}{2}} \int_{\Omega_\varepsilon(\xi)} \left(\frac{u_\varepsilon(x + \varepsilon\xi) - u_\varepsilon(x)}{\varepsilon\xi} \right)^2 dx \, d\xi$$

$$+ 2 \int_{\frac{1}{2}}^1 \int_{\Omega_\varepsilon(2\xi)} \left(f_1 \left(\frac{u_\varepsilon(x + \varepsilon\xi) - u_\varepsilon(x)}{\varepsilon\xi} \right) \right.$$

$$+ 2 f_2 \left(\frac{u_\varepsilon(x + 2\varepsilon\xi) - u_\varepsilon(x)}{2\varepsilon\xi} \right) \Bigg) dx \, d\xi$$

$$\geq 2 \int_0^{\frac{1}{2}} \int_{\Omega_\varepsilon(\xi)} \left(\frac{u_\varepsilon(x + \varepsilon\xi) - u_\varepsilon(x)}{\varepsilon\xi} \right)^2 dx \, d\xi$$

$$+ 2 \int_{\frac{1}{2}}^1 \sum_{l=0}^{\lfloor 1/\varepsilon\xi \rfloor - 3} \int_{l\varepsilon\xi}^{(l+1)\varepsilon\xi} \left(f_1 \left(\frac{u_\varepsilon(x + \varepsilon\xi) - u_\varepsilon(x)}{\varepsilon\xi} \right) \right.$$

$$+ 2 f_2 \left(\frac{u_\varepsilon(x + 2\varepsilon\xi) - u_\varepsilon(x)}{2\varepsilon\xi} \right) \Bigg) dx \, d\xi$$

Through the change of variable $x = \varepsilon\xi(l + t)$, $t \in (0, 1)$, we then get for $r < 1$ and ε small enough

$$F_\varepsilon(u_\varepsilon) \geq 2 \int_0^{\frac{1}{2}} \int_{\Omega_\varepsilon(\xi)} \left(\frac{u_\varepsilon(x + \varepsilon\xi) - u_\varepsilon(x)}{\varepsilon\xi} \right)^2 dx \, d\xi$$

$$+ 2 \int_{\frac{1}{2}}^1 \int_0^1 \sum_{l=0}^{\lfloor 1/\varepsilon\xi \rfloor - 3} \varepsilon\xi \left(f_1 \left(\frac{u_\varepsilon(\varepsilon\xi(l + t + 1)) - u_\varepsilon(\varepsilon\xi(l + t))}{\varepsilon\xi} \right) \right.$$

$$+ 2 f_2 \left(\frac{u_\varepsilon(\varepsilon\xi(l + t + 2)) - u_\varepsilon(\varepsilon\xi(l + t))}{2\varepsilon\xi} \right) \Bigg) dt \, d\xi$$

$$\geq 2 \int_0^{\frac{1}{2}} \int_{\Omega_\varepsilon(\xi)} \left(\frac{u_\varepsilon(x + \varepsilon\xi) - u_\varepsilon(x)}{\varepsilon\xi} \right)^2 dx \, d\xi$$

$$+ 2 \int_{\frac{1}{2}}^{1} \int_{0}^{1} F_{\varepsilon\xi}^{\text{disc}}((u_{\varepsilon})^{\varepsilon\xi, t}, (0, r)) \, dt \, d\xi$$

$$=: I_1^{\varepsilon} + I_2^{\varepsilon}.$$

We can apply now Theorem 3.1. Since $u_{\varepsilon} \to zx$ in $L^2(0, 1)$ we deduce that $\liminf_{\varepsilon \to 0} I_1^{\varepsilon} \geq z^2$. By (6.26), (6.27) and Fatou's Lemma, we get that $\liminf_{\varepsilon \to 0} I_2^{\varepsilon} \geq r(g_0)^{**}(z)$ for every $r < 1$. Hence, we conclude that

$$f_{\text{hom}}(z) \geq z^2 + (g_0)^{**}(z) > \overline{f}^{**}(z)$$

whenever $z \in (-1, 0) \cup (0, 1)$.

Example 6.3 ($f_{\text{hom}}(M) < Q\overline{f}(M)$) Let F_{ε} be defined by (2.1), with $d = m = 1$, $\Omega = (0, 1)$ and $f_{\varepsilon}(x, \xi, z) = f(\xi, z)$, where

$$f(\xi, z) := \begin{cases} f_1\left(\dfrac{z}{\xi}\right) & \text{if } 0 < |\xi| \leq 1 \\ f_2\left(\dfrac{z}{\xi}\right) & \text{if } k \leq |\xi| \leq k + 1 \\ 0 & \text{otherwise,} \end{cases}$$

with k to be chosen and

$$f_1(z) = (|z| - 1)^2, \quad f_2(z) = z^2.$$

Note that

$$\overline{f}(z) = 2(|z| - 1)^2 + 2z^2;$$

hence,

$$Q\overline{f}(z) = (\overline{f})^{**}(z) = \begin{cases} 1 & \text{if } |z| \leq 1/2 \\ 2(|z| - 1)^2 + 2z^2 & \text{if } |z| > 1/2. \end{cases}$$

We now show that for $k > \sqrt{6}$

$$f_{\text{hom}}(0) < 1 = (\overline{f})^{**}(0).$$

To this end, let $u : \mathbb{R} \to \mathbb{R}$ be 2-periodic and such that

$$u(x) = \begin{cases} x & \text{if } 0 \leq x \leq 1 \\ 2 - x & \text{if } 1 \leq x \leq 2 \end{cases}$$

and set $u_\varepsilon(x) := \varepsilon u(\varepsilon^{-1}x)$. Note that $u_\varepsilon \rightharpoonup 0$ weakly in $H^1(0, 1)$, hence

$$f_{\text{hom}}(0) \le \limsup_{\varepsilon \to 0} F_\varepsilon(u_\varepsilon).$$

We get that

$$
\begin{aligned}
F_\varepsilon(u_\varepsilon) = 2 \sum_{j=0,\, j \text{ even}}^{\lfloor 1/\varepsilon \rfloor} \Big(& \int_0^1 \int_{j\varepsilon}^{(j+2)\varepsilon} f_1\Big(\frac{u_\varepsilon(x + \varepsilon\xi) - u_\varepsilon(x)}{\varepsilon\xi}\Big) dx\, d\xi \\
& + 2 \int_k^{k+1} \int_{j\varepsilon}^{(j+2)\varepsilon} f_2\Big(\frac{u_\varepsilon(x + \varepsilon\xi) - u_\varepsilon(x)}{\varepsilon\xi}\Big) dx\, d\xi \Big) + o(1),
\end{aligned}
$$

thus

$$
\begin{aligned}
\lim_{\varepsilon \to 0} F_\varepsilon(u_\varepsilon) = 2 & \int_0^1 \int_0^2 f_1\Big(\frac{u(x + \xi) - u(x)}{\xi}\Big) dx\, d\xi \\
& + 2 \int_k^{k+1} \int_0^2 f_2\Big(\frac{u(x + \xi) - u(x)}{\xi}\Big) dx\, d\xi =: S_1 + S_2.
\end{aligned}
$$

A direct computation shows that

$$S_1 = \frac{1}{2} \int_{-1}^1 f_1(z)\, dz = \frac{1}{3},$$

while $S_2 \le \frac{2}{k^2}$. Hence

$$f_{\text{hom}}(0) \le \frac{1}{3} + \frac{4}{k^2} < 1$$

whenever $k > \sqrt{6}$.

In the next example we show that in the vector case f_{hom} may be a non convex quasi-convex function.

Example 6.4 (A Quasiconvex Non-convex Limit Energy Density) Let F_ε be defined by (2.1), with $d = m = 2$, $\Omega = (0, 1)^2$, $f_\varepsilon(x, \xi, z) \equiv 0$ if $|\xi| > 1$ and for every $\xi \in B_1 \setminus \{0\}$

$$f_\varepsilon(x, \xi, z) = f(\xi, z) := \begin{cases} 1 + \left(\dfrac{|z|}{|\xi|}\right)^p & \text{if } z \ne \pm\xi \\ 0 & \text{if } z = \pm\xi. \end{cases}$$

By testing the minimum problem defining f_{hom} with the identity function and its opposite, we immediately get that

$$f_{\text{hom}}(I) = f_{\text{hom}}(-I) = 0,$$

where I is the identity matrix in $\mathbb{R}^{2\times 2}$. We now show that $f_{\text{hom}}(0) > 0$, from which we get that f_{hom} is not convex. The proof relies on a suitable lower bound of F_ε with discrete energies, for which an analogous result was proven in [2]. Let us introduce some notation. Given $\xi \in \mathbb{R}^2 \setminus \{0\}$, let $\mathcal{L}_\xi$ be the lattice in $\mathbb{R}^2$ defined by

$$\mathcal{L}_\xi = \mathbb{Z}\xi \oplus \mathbb{Z}\xi^\perp,$$

where $\xi^\perp := (-\xi_2, \xi_1)$. We now introduce the functionals $F_\varepsilon^{\xi,\text{disc}}(u, A)$ defined on discrete functions $u : \varepsilon\mathcal{L}_\xi \cap (0, 1)^2 \to \mathbb{R}^2$ and localized on each open set A in $\mathbb{R}^2$, as

$$F_\varepsilon^{\xi,\text{disc}}(u, A) = \sum_{\xi' \in \{\xi, \xi^\perp, \xi+\xi^\perp\}} \sum_{k \in A^{\varepsilon,\xi'}} (\varepsilon|\xi|)^2 f\left(\xi', \frac{u(k + \varepsilon\xi') - u(k)}{\varepsilon}\right),$$

where, for $\xi' \in \mathcal{L}_\xi \setminus 0$, we have set

$$A^{\varepsilon,\xi'} := \varepsilon\mathcal{L}_\xi \cap A \cap (A - \varepsilon\xi').$$

The discrete functions u can be regarded as L^p functions by identifying them with their piecewise interpolation on the elementary cells of the lattice $\varepsilon\mathcal{L}_\xi$, that is, with a little abuse of notation,

$$u(x) = u(k) \text{ if } x \in k + \varepsilon Q^\xi, \ k \in \varepsilon\mathcal{L}_\xi \cap (0, 1)^2,$$

where

$$Q^\xi := [0, 1)\xi \oplus [0, 1)\xi^\perp.$$

In [2], it was proved that for any open set $A \subset \mathbb{R}^2$ with Lipschitz boundary

$$\Gamma\text{-}\lim_{\varepsilon\to 0} F_\varepsilon^{e_1,\text{disc}}(u, A) = \int_A f_{\text{hom}}^{\text{disc}}(\nabla u)\, dx, \quad u \in W^{1,p}(A; \mathbb{R}^2),$$

where the Γ-limit is performed with respect to the L^p-convergence, and that $f_{\text{hom}}^{\text{disc}}(0) > 0$ (see [2] Theorem 7.1). Then, for any $\xi \in \mathbb{R}^2 \setminus \{0\}$

$$\Gamma\text{-}\lim_{\varepsilon\to 0} F_\varepsilon^{\xi,\text{disc}}(u, A) = \int_A f_{\text{hom}}^{\text{disc}}(R_\xi^T \nabla u R_\xi)\, dx, \quad u \in W^{1,p}(A; \mathbb{R}^2), \qquad (6.28)$$

where R_ξ is the rotation in $\mathbb{R}^{2\times 2}$ such that $R_\xi e_1 = \frac{\xi}{|\xi|}$.

Regarding the convergence of proper discrete interpolation of L^1 functions, in [3] the following result was proved. Given $u \in L^1_{loc}(\mathbb{R}^2; \mathbb{R}^2)$, set for $y \in Q^\xi$

$$T^{\varepsilon,\xi}_y u(x) := u(\varepsilon y + \varepsilon \lfloor x/\varepsilon \rfloor_\xi),$$

where

$$\lfloor z \rfloor_\xi := \lfloor z \cdot \xi \rfloor \xi + \lfloor z \cdot \xi^\perp \rfloor \xi^\perp.$$

If $u_\varepsilon \to u$ in $L^1_{loc}(\mathbb{R}^2; \mathbb{R}^2)$, then

$$T^{\varepsilon,\xi}_y u_\varepsilon \to u \text{ in } L^1_{loc}(\mathbb{R}^2; \mathbb{R}^2) \text{ for a.e. } y \in Q^\xi \qquad (6.29)$$

(see [3, Lemma 2.11]).

We are now in a position to prove the claim. Let $u_\varepsilon \to 0$ in $L^p((0, 1)^2; \mathbb{R}^2)$ such that

$$\lim_{\varepsilon \to 0} F_\varepsilon(u_\varepsilon) = f_{\hom}(0).$$

We easily get the following lower bound for a square $Q' \Subset \Omega$ and ε small enough

$$F_\varepsilon(u_\varepsilon) \geq \frac{1}{3} \int_{B_{\frac{\sqrt{2}}{2}}} \int_{Q'} \sum_{\xi' \in \{\xi, \xi^\perp, \xi + \xi^\perp\}} f\left(\xi', \frac{u_\varepsilon(x + \varepsilon\xi') - u_\varepsilon(x)}{\varepsilon}\right) dx \, d\xi$$

$$\geq \frac{1}{3} \int_{B_{\frac{\sqrt{2}}{2}}} \sum_{k \in K^\xi_\varepsilon} \int_{k + \varepsilon Q^\xi} \sum_{\xi' \in \{\xi, \xi^\perp, \xi + \xi^\perp\}} f\left(\xi', \frac{u_\varepsilon(x + \varepsilon\xi') - u_\varepsilon(x)}{\varepsilon}\right) dx \, d\xi,$$

where

$$K^\xi_\varepsilon := \{k \in \varepsilon \mathcal{L}_\xi : k + \varepsilon Q^\xi \subset Q'\}.$$

Through the change of variable $x = k + \varepsilon y$, $y \in Q^\xi$, we then get for a square $Q'' \Subset Q'$ and ε small enough

$$F_\varepsilon(u_\varepsilon) \geq$$

$$\frac{1}{3} \int_{B_{\frac{\sqrt{2}}{2}}} \int_{Q^\xi} \sum_{k \in K^\xi_\varepsilon} \sum_{\xi' \in \{\xi, \xi^\perp, \xi + \xi^\perp\}} \varepsilon^2 f\left(\xi', \frac{u_\varepsilon(k + \varepsilon y + \varepsilon\xi') - u_\varepsilon(k + \varepsilon y)}{\varepsilon}\right) dy \, d\xi$$

$$\geq \frac{1}{3} \int_{B_{\frac{\sqrt{2}}{2}}} \frac{1}{|\xi|^2} \int_{Q^\xi} F^{\xi,\text{disc}}_\varepsilon(T^{\varepsilon,\xi}_y u_\varepsilon, Q'') \, dy \, d\xi \, .$$

By (6.28), (6.29), Fatou's Lemma and the arbitrariness of Q'', we conclude that

$$f_{\text{hom}}(0) \geq \frac{\pi}{6} f_{\text{hom}}^{\text{disc}}(0) > 0.$$

6.4   An Extension Lemma from Periodic Lipschitz Domains

We prove the existence of an extension operator for non-local functionals defined on general connected domains. For the reader convenience we recall the definition of Lipschitz boundary that will be used in the proof of Lemma 6.1.

Definition 6.1 An open set $E \subset \mathbb{R}^n$ has *Lipschitz boundary at* $x \in \partial E$ if ∂E is locally the graph of a Lipschitz function, in the sense that there exist a coordinate system $(y_1, \ldots, y_d)$ obtained by a translation and a rotation, a Lipschitz function Φ of $d-1$ variables, and an open cube U_x in the y-coordinates, centred at x, such that $E \cap U_x = \{y : y_d < \Phi(y_1, \ldots, y_{d-1})\}$ and that ∂E splits U_x into two connected sets, $E \cap U_x$ and $U_x \setminus \overline{E}$. If this property holds for every $x \in \partial E$ with the same Lipschitz constant L and side length ρ of U_x, we say that E has *Lipschitz boundary*.

Throughout the following two sections, for the sake of brevity, we will adopt the following notation to indicate a neighbourhood of thickness $r > 0$ of the diagonal in $\mathbb{R}^d \times \mathbb{R}^d$,

$$D_r = \{(x, y) \in \mathbb{R}^d \times \mathbb{R}^d : |x - y| < r\}.$$

Theorem 6.3 (Extension Theorem) *Let E be a periodic open subset of $\mathbb{R}^d$ with Lipschitz boundary and let Ω be a bounded open subset of $\mathbb{R}^d$. Then, there exist $R = R(E) > 0$ and $k_0 > 0$ such that for all $\varepsilon > 0$ there exists a linear and continuous extension operator $T_\varepsilon : L^p(\Omega \cap \varepsilon E) \to L^p(\Omega)$ such that for all $r > 0$ and for all $u \in L^p(\Omega \cap \varepsilon E)$,*

$$T_\varepsilon u = u \quad \text{almost everywhere in} \quad \Omega \cap \varepsilon E, \tag{6.30}$$

$$\int_{\Omega(\varepsilon k_0)} |T_\varepsilon u|^p \, dx \leq c_1 \int_{\Omega \cap \varepsilon E} |u|^p \, dx, \tag{6.31}$$

$$\int_{(\Omega(\varepsilon k_0) \times \Omega(\varepsilon k_0)) \cap D_{\varepsilon R}} |T_\varepsilon u(x) - T_\varepsilon u(y)|^p \, dx \, dy$$

$$\leq c_2(r) \int_{(\Omega(\varepsilon k_0) \times \Omega(\varepsilon k_0)) \cap D_{\varepsilon r}} |u(x) - u(y)|^p \, dx \, dy, \tag{6.32}$$

where we use notation

$$\Omega(\lambda) := \{x \in \Omega : \text{dist}(x, \partial\Omega) > \lambda\}. \tag{6.33}$$

The positive constants c_1 and c_2 depend on E and d and, in addition, c_2 depends also on r, but both are independent of ε.

We will prove the theorem in a special case, when

$$E = \mathbb{R}^d \setminus (\mathbb{Z}^d + \mathbf{K}),$$

where $\mathbf{K}$ is a compact set with C^2-boundary such that $(i + \mathbf{K}) \cap (j + \mathbf{K}) = \emptyset$ if $i, j \in \mathbb{Z}^d$ and $i \neq j$. The case with separate compact perforations contains the relevant arguments due to the non-local form of the functionals separated from the issues on perforated domains that are common with local functionals. The case of a general E is more technical, involving a partition-of-unity argument as in the case of local integral functionals on perforated domains treated by Acerbi et al. [1] (for details see [6]).

Lemma 6.1 *Let A be a connected bounded set with Lipschitz boundary. For any $r > 0$ there exists a constant $c_r > 0$ such that the following inequality holds*

$$\int_{A \times A} |u(\eta) - u(\xi)|^p \, d\xi \, d\eta \leq c_r \int_{(A \times A) \cap D_r} |u(\eta) - u(\xi)|^p \, d\xi \, d\eta. \tag{6.34}$$

If L, ρ are constants as in Definition 6.1 then the constant c_r depends on A only through its diameter and such constants.

Proof Since for any function u the integral on the right-hand side of (6.34) is an increasing function of r, it is sufficient to prove (6.34) for r positive and small enough.

Since A has a Lipschitz boundary and is connected, with fixed $r > 0$ there exists $r_1 \in (0, \frac{1}{2}r)$ and $\nu \in (0, 1/2]$ that only depends on the Lipschitz constant of A such that for any two points η', $\eta'' \in A$ there is a discrete path from η' to η''; i.e., a set of points $\eta' = \eta_0, \eta_1, \ldots, \eta_N, \eta'' = \eta_{N+1}$, that possesses the following properties:

(a) $|\eta_{j+1} - \eta_j| \leq r_1$ for $j = 0, 1, \ldots, N$;
(b) for all $j = 1, \ldots, N$ the ball $\overline{B}_{\nu r_1}(\eta_j)$ is contained in A;
(c) there exists $\bar{N} = \bar{N}(r_1, \text{diam}(A))$ such that $N \leq \bar{N}$ for all $\eta', \eta'' \in A$.

Indeed, since A is a bounded set with Lipschitz boundary, it has a Lipschitz continuous boundary. Then there exists a constant $r_2 = r_2(L, \rho, r) > 0$ such that $r_2 < \frac{1}{2}r$, $r_2 < \frac{1}{2d}\rho \min(1, L)$, and the set $A(r_2)$, defined as in (6.33), is connected. We choose $r_1 = \frac{r_2}{8(L+1)}$ and denote $Z_A = \{z \in \frac{r_1}{\sqrt{d}}\mathbb{Z}^d : z \in A_{r_2}\}$. By construction $B_{r_1}(x) \subset A$ for any $x \in A(r_2)$, and for any z_1 and z_2 in Z_A there exists a path $z_1 = \eta_1, \ldots, \eta_N = z_2$ in Z_A such that $|\eta_{j+1} - \eta_j| \leq r_1$ and $N \leq \left(\frac{\text{diam}(A)}{r_1}\right)^d$. Also, by construction, for any $x \in A \setminus A(r_2)$ there exists a path $x = \tilde{\eta}_0, \ldots, \tilde{\eta}_{\tilde{N}}$ such that $\tilde{\eta}_{\tilde{N}} \in Z_A$, $|\tilde{\eta}_{j+1} - \tilde{\eta}_j| \leq r_1$, $\tilde{N} \leq 16(L+1)d$, and $B_{\frac{r_1}{2(L+1)}}(\tilde{\eta}_j) \subset A$ for all $j = 1, \ldots, \tilde{N}$. This implies the existence of a path that has properties (a)–(c).

Writing

$$u(\xi_0) - u(\xi_{N+1}) = u(\xi_0) - u(\xi_1) + u(\xi_1) - \ldots - u(\xi_N) + u(\xi_N) - u(\xi_{N+1}),$$

where ξ_j denotes a point in $B_{vr_1}(\eta_j)$ for $1 \leq j \leq N$, we get

$$\int_{(A \cap B_{vr_1}(\eta')) \times (A \cap B_{vr_1}(\eta''))} |u(\xi_0) - u(\xi_{N+1})|^p \, d\xi_0 d\xi_{N+1}$$

$$= \frac{C}{(vr_1)^{dN}} \int_{B_{vr_1}(\eta_1)} \cdots \int_{B_{vr_1}(\eta_N)} \int_{(A \cap B_{vr_1}(\eta')) \times (A \cap B_{vr_1}(\eta''))} \Big| u(\xi) - u(\xi_1)$$

$$+ u(\xi_1) - \ldots - u(\xi_N) + u(\xi_N) - u(\eta) \Big|^p \, d\xi_0 \, d\xi_{N+1} \, d\xi_N \ldots d\xi_1$$

$$\leq C \frac{(N+1)^{p-1}}{(vr_1)^{dN}} \int_{A \cap B_{vr_1}(\eta_0)} \cdots \int_{A \cap B_{vr_1}(\eta_{N+1})} \sum_{j=1}^{N+1} |u(\xi_j) - u(\xi_{j-1})|^p d\xi_{N+1} \ldots d\xi_0$$

$$= C (N+1)^{p-1} \sum_{j=1}^{N+1} \int_{(A \cap B_{vr_1}(\eta_j)) \times (A \cap B_{vr_1}(\eta_{j-1}))} |u(\xi_j) - u(\xi_{j-1})|^p d\xi_j \, d\xi_{j-1}$$

$$\leq C(\tilde{N}+1)^p \int_{(A \times A) \cap D_r} |u(\eta) - u(\xi)|^p \, d\xi \, d\eta.$$

Covering A with a finite number of balls of radius vr_1 and summing up the last inequality over all pairs of these balls gives the desired estimate (6.34). □

Proof (of Theorem 6.3 for Compact Perforations) We apply our arguments separately to each connected component of $\mathbb{R}^d \setminus E$. With fixed $\tau > 0$ chosen below we consider a connected component $\mathbf{K}$ of $\mathbb{R}^d \setminus E$, and set

$$A := \{\xi \in \mathbb{R}^d \setminus \mathbf{K} : \text{dist}(\xi, \partial \mathbf{K}) < \tau\} \text{ and } A^\star := \{\xi \in \mathbf{K} : \text{dist}(\xi, \partial \mathbf{K}) < \tau\}.$$

Since $\mathbf{K}$ is bounded and C^2, we may fix $\tau > 0$ small enough and an invertible mapping $\mathcal{R}$ from A to $A^\star$ such that

$$\frac{1}{2} |\mathcal{R}(\xi') - \mathcal{R}(\xi'')| \leq |\xi' - \xi''| \leq 2|\mathcal{R}(\xi') - \mathcal{R}(\xi'')|$$

for all ξ', $\xi'' \in A$. Slightly abusing the notation we call this mapping a *reflection*. In what follows for the sake of brevity we use the notation $\xi_\mathcal{R} = \mathcal{R}^{-1}(\xi)$ for $\xi \in A^\star$.

We set

$$\bar{u}_A = \frac{1}{|A^\star|} \int_{A^\star} u(\xi_R)\, d\xi.$$

Let φ be a C^∞ function such that $0 \le \varphi \le 1$, $\varphi = 1$ in A and in a neighbourhood of $\partial \mathbf{K}$, $\varphi = 0$ in a neighbourhood of $\partial A^\star \setminus \partial \mathbf{K}$.

We define $v(\xi)$ as follows

$$v(\xi) = \begin{cases} u(\xi) & \text{if } \xi \in A \\ \varphi(\xi)u(\xi_R) + (1 - \varphi(\xi))\bar{u}_A & \text{if } \xi \in A^\star \\ \bar{u}_A & \text{if } \xi \in \mathbf{K} \setminus A^\star. \end{cases}$$

Letting $k_0 = \operatorname{diam}(Q_1) = \sqrt{d}$ and $R = \min(r, k_0, \tau)$, we have

$$\int_{(A \times A) \cap D_R} |v(\eta) - v(\xi)|^p \, d\xi d\eta = \int_{(A \times A) \cap D_R} |u(\eta) - u(\xi)|^p \, d\xi d\eta \qquad (6.35)$$

and

$$\int_{(A \times A^\star) \cap D_R} |v(\eta) - v(\xi)|^p \, d\xi d\eta \le \int_{A \times A} |u(\eta) - u(\zeta)|^p \left| \frac{\partial \mathcal{R}(\zeta)}{\partial \zeta} \right| d\zeta d\eta$$

$$\le C_{\mathcal{R}} \int_{A \times A} |u(\eta) - u(\zeta)|^p \, d\zeta d\eta. \qquad (6.36)$$

Here we have used the fact that the Jacobian $\left| \frac{\partial \mathcal{R}(\zeta)}{\partial \zeta} \right|$ is a bounded function: $\left| \frac{\partial \mathcal{R}(\zeta)}{\partial \zeta} \right| \le C_{\mathcal{R}}$.

Next, taking into account the relation

$$v(\xi) - v(\eta) = (\varphi(\xi) - \varphi(\eta))(u(\xi_R) - \bar{u}_A) + \varphi(\eta)(u(\xi_R) - u(\eta_R)) \text{ if } \eta \in A^\star,\ \xi \in A^\star$$

we obtain

$$\int_{(A^\star \times A^\star) \cap D_R} |v(\eta) - v(\xi)|^p \, d\xi d\eta$$

$$\le \int_{A^\star \times A^\star} |\bar{u}_A - u(\xi_R)|^p \, d\xi d\eta + \int_{A^\star \times A^\star} |u(\eta_R) - u(\xi_R)|^p \, d\xi d\eta.$$

Since $\bar{u}_A$ is the average of the function $u(\xi_R)$ over $A^\star$, then

$$\int_{A^\star \times A^\star} |\bar{u}_A - u(\xi_R)|^p \, d\xi d\eta \le \int_{A^\star \times A^\star} |u(\eta_R) - u(\xi_R)|^p \, d\xi d\eta.$$

This yields

$$\int_{(A^\star \times A^\star) \cap D_R} |v(\eta) - v(\xi)|^p \, d\xi d\eta \leq 2 \, C_R^2 \int_{A \times A} |u(\eta) - u(\xi)|^p \, d\xi d\eta. \qquad (6.37)$$

Finally,

$$\int_{(K \setminus A^\star) \cap D_R} \int_{A^\star} |v(\eta) - v(\xi)|^p \, d\xi d\eta$$

$$\leq \int_{K \setminus A^\star} \int_{A^\star} |\varphi(\xi)|^p |\bar{u}_A - u(\xi_R)|^p \, d\xi \, d\eta$$

$$\leq |K \setminus A^\star| \int_{A^\star} |\bar{u}_A - u(\xi_R)|^p \, d\xi$$

$$\leq C_R \frac{|K \setminus A^\star|}{|A^\star|} \int_{A \times A} |u(\eta) - u(\xi)|^p \, d\xi d\eta.$$

Combining the last inequality with (6.35), (6.36) and (6.37) we conclude that

$$\int_{((K \cup A) \times (K \cup A)) \cap D_R} |v(\eta) - v(\xi)|^p \, d\xi d\eta \leq C \int_{A \times A} |u(\eta) - u(\xi)|^p \, d\xi d\eta \qquad (6.38)$$

We may now apply Lemma 6.1. By (6.34) we obtain

$$\int_{((K \cup A) \times (K \cup A)) \cap D_R} |v(\eta) - v(\xi)|^p \, d\xi d\eta$$

$$\leq C \int_{(A \times A) \cap D_R} |u(\eta) - u(\xi)|^p \, d\xi d\eta.$$

After rescaling, this inequality reads

$$\int_{(\varepsilon(K \cup A) \times \varepsilon(K \cup A)) \cap D_{\varepsilon r}} |v(\eta) - v(\xi)|^p \, d\xi d\eta$$

$$\leq C_1 \int_{(\varepsilon A \times \varepsilon A) \cap D_{\varepsilon r}} |u(\eta) - u(\xi)|^p \, d\xi d\eta \qquad (6.39)$$

Summing up the last inequality over all the inclusions in $\Omega(k_0 \varepsilon)$, choosing $T_\varepsilon u = v$ we obtain (6.32). Condition (6.30) and inequality (6.31) are straightforward consequences of the definition of v. $\square$

Corollary 6.1 *Let u_ε be a family of functions in $L^p(\Omega \cap \varepsilon E)$ such that there exists $c > 0$ and $r > 0$ such that $\|u_\varepsilon\|_{L^p(\Omega \cap \varepsilon E)} \leq c$ and*

$$\int_{B_r} \int_{(\Omega \cap \varepsilon E)_\varepsilon(\xi)} \left| \frac{u_\varepsilon(x + \varepsilon\xi) - u_\varepsilon(x)}{\varepsilon} \right|^p dx \, d\xi \leq c, \tag{6.40}$$

for all $\varepsilon > 0$, with $(\Omega \cap \varepsilon E)_\varepsilon(\xi) = \{x \in \Omega \cap \varepsilon E : x + \varepsilon\xi \in \Omega \cap \varepsilon E\}$. Then, for any sequence $\varepsilon_j \to 0$ as $j \to +\infty$, and for any open subset $\Omega' \Subset \Omega$ the set $\{T_{\varepsilon_j} u_{\varepsilon_j}\}_j$ is relatively compact in $L^p(\Omega')$ and every its limit point is in $W^{1,p}(\Omega)$.

Proof Let u_ε be such that $\|u_\varepsilon\|_{L^p(\Omega \cap \varepsilon E)} \leq c$ and (6.40) hold for every $\varepsilon > 0$. From Theorem 6.3, the extended functions $T_\varepsilon u_\varepsilon$ satisfy the estimates

$$\int_{\Omega(\varepsilon k_0)} |T_\varepsilon u_\varepsilon|^p \, dx \leq c \tag{6.41}$$

and

$$\frac{1}{\varepsilon^{d+p}} \int_{(\Omega(\varepsilon k_0) \times (\Omega(\varepsilon k_0))) \cap D_{\varepsilon R}} |T_\varepsilon u_\varepsilon(y) - T_\varepsilon u_\varepsilon(x)|^p \, dy \, dx$$

$$\leq c_2(r) \int_{B_r} \int_{(\Omega \cap E)_\varepsilon(\xi)} \left| \frac{u_\varepsilon(x + \varepsilon\xi) - u_\varepsilon(x)}{\varepsilon} \right|^p dx \, d\xi \leq c,$$

for some $R > 0$ independent of ε. The latter, after the change of variables $y = x + \varepsilon\xi$, is equivalent to

$$\int_{\Omega(\varepsilon k_0)} \int_{B_R} \left| \frac{T_\varepsilon u_\varepsilon(x + \varepsilon\xi) - T_\varepsilon u_\varepsilon(x)}{\varepsilon} \right|^p d\xi \, dx \leq c,$$

which corresponds to

$$\int_{\Omega(\varepsilon k_0)} \int_{B_R} \left| \frac{w_\varepsilon(x + \varepsilon\xi) - w_\varepsilon(x)}{\varepsilon} \right|^p d\xi \, dx \leq c \tag{6.42}$$

for $w_\varepsilon = T_\varepsilon u_\varepsilon$. Using Theorem 4.2 for w_ε and (6.41), (6.42), we can conclude that for any sequence $\varepsilon_j \to 0$ as $j \to +\infty$, and for any open subset $\Omega' \subset\subset \Omega$, $T_\varepsilon u_{\varepsilon_j}$ is relatively compact in $L^p(\Omega')$ and every its limit point is in $W^{1,p}(\Omega)$. $\qquad \square$

6.5 Homogenization on Perforated Domains

In this section we present an application of the Extension Theorem 6.3 to the homogenization of non-local functional. Specifically, we consider a periodic integrand $h : \mathbb{R}^d \times \mathbb{R}^d \times \mathbb{R}^m \to [0, +\infty)$; i.e., a Borel function such that $h(\cdot, \xi, z)$

is $[0, 1]^d$-periodic for all $\xi \in \mathbb{R}^d$ and $z \in \mathbb{R}^m$ and satisfies the following growth conditions: there exist positive constants c_0, c_1, r_0 and non-negative function $\psi :$ $\mathbb{R}^d \to [0, +\infty)$ such that

$$h(x, \xi, z) \le \psi(\xi)(|z|^p + 1) \tag{6.43}$$

$$h(x, \xi, z) \ge c_0(|z|^p - 1) \quad \text{for all } |\xi| \le r_0 \tag{6.44}$$

with

$$\int_{\mathbb{R}^d} \psi(\xi)|\xi|^p \, d\xi \le c_1. \tag{6.45}$$

Let $\Omega \subset \mathbb{R}^d$ be an open set with Lipschitz boundary. For any $\varepsilon > 0$, we introduce the non-local functional $H_\varepsilon : L^p(\Omega; \mathbb{R}^m) \to [0, +\infty]$ defined as

$$H_\varepsilon(u) = \int_{\mathbb{R}^d} \int_{(\Omega \cap \varepsilon E)_\varepsilon(\xi)} h\left(\frac{x}{\varepsilon}, \xi, \frac{u(x + \varepsilon \xi) - u(x)}{\varepsilon}\right) dx \, d\xi. \tag{6.46}$$

Note that the integration in (6.46) is performed for x, ξ such that both x and $x + \varepsilon \xi$ belong to the perforated domain $\Omega \cap \varepsilon E$. Conditions (6.43)–(6.45) guarantee that functionals H_ε are estimated from above and below as in Sect. 2.3.

Thanks to Corollary 6.1, our functionals H_ε are equi-coercive with respect to the $L^p_{\text{loc}}(\Omega)$-convergence upon identifying functions with their extensions from the perforated domain. More precisely, from each sequence $\{u_\varepsilon\}$ with equi-bounded energy $H_\varepsilon(u_\varepsilon)$ we can extract a subsequence such that the corresponding extensions converge in L^p_{loc} to some limit $u \in W^{1,p}(\Omega)$. This is implied by Corollary 6.1 applied with $r = r_0$ to each component of the vector-valued functions u_ε, upon noting that (6.44) implies (6.40).

We now may state the homogenization result for the functional H_ε with respect to the $L^p_{\text{loc}}(\Omega; \mathbb{R}^m)$ convergence.

Theorem 6.4 (Homogenization on Perforated Domains) *The functionals H_ε defined by (6.46) Γ-converge with respect to $L^p_{\text{loc}}(\Omega; \mathbb{R}^m)$-convergence to the functional*

$$H_{\text{hom}}(u) = \begin{cases} \int_\Omega h_{\text{hom}}(\nabla u(x)) \, dx & \text{if } u \in W^{1,p}(\Omega; \mathbb{R}^m) \\ +\infty & \text{otherwise,} \end{cases} \tag{6.47}$$

with h_{hom} satisfying the asymptotic formula

$$h_{\text{hom}}(\Xi) = \lim_{T \to +\infty} \frac{1}{T^d} \inf \left\{ \int_{(0,T)^d \cap E} \int_{(0,T)^d \cap E} h(x, y - x, v(y) - v(x)) \, dx \, dy : \right.$$

$$\left. v(x) = \Xi x \text{ if } \text{dist}(x, \partial(0, T)^d) < k_0 \right\} \tag{6.48}$$

for all $\Xi \in \mathbb{R}^{m \times d}$ and k_0 as in Theorem 6.3. Furthermore, if h is convex in the third variable, the following cell-problem formula holds

$$h_{\text{hom}}(\Xi) = \inf\left\{\int_{(0,1)^d \cap E}\int_E h(x, y - x, v(y) - v(x))\, dx\, dy\right.$$

$$\left. : v(x) - \Xi x \text{ is } 1\text{-periodic}\right\}. \qquad (6.49)$$

Proof We will prove Theorem 6.4 reducing to Theorem 6.1 (i.e., to $E = \mathbb{R}^d$) by a perturbation argument. For every $\delta \geq 0$ we set

$$h^\delta(x, \xi, z) = \chi_E(x)\chi_E(x + \xi)\, h(x, \xi, z) + \delta\chi_{B_{R_0}}(\xi)|z|^p,$$

where $R_0 > 0$ is fixed but arbitrary, and

$$H_\varepsilon^\delta(u) = \int_{\mathbb{R}^d}\int_{\Omega_\varepsilon(\xi)} h^\delta\left(\frac{x}{\varepsilon}, \xi, \frac{u(x + \varepsilon\xi) - u(x)}{\varepsilon}\right) dx\, d\xi$$

is defined for $u \in L^p(\Omega; \mathbb{R}^m)$. Note that $H_\varepsilon^\delta \geq H_\varepsilon$, and for $\delta = 0$ we have $H_\varepsilon^0 = H_\varepsilon$. In the following, for any open set A and $\delta \geq 0$, we also consider the 'localized' functionals

$$H_\varepsilon^\delta(v, A) = \int_{\mathbb{R}^d}\int_{A_\varepsilon(\xi)} h^\delta\left(\frac{x}{\varepsilon}, \xi, \frac{u(x + \varepsilon\xi) - u(x)}{\varepsilon}\right) dx\, d\xi.$$

If $\delta = 0$ we write $H_\varepsilon(v, A)$ in the place of $H_\varepsilon^0(v, A)$.

Theorem 6.1 ensures that for all $\delta > 0$ there exists the Γ-limit

$$H_{\text{hom}}^\delta(u) = \Gamma\text{-}\lim_{\varepsilon \to 0} H_\varepsilon^\delta(u)$$

with domain $W^{1,p}(\Omega; \mathbb{R}^m)$, on which it is represented as

$$H_{\text{hom}}^\delta(u) = \int_\Omega h_{\text{hom}}^\delta(\nabla u)\, dx.$$

The energy density h_{hom}^δ satisfies

$$h_{\text{hom}}^\delta(\Xi) = \lim_{T \to +\infty}\frac{1}{T^d}\inf\left\{\int_{(0,T)^d}\int_{(0,T)^d} h^\delta(x, y - x, v(y) - v(x))\, dx\, dy :\right.$$

$$\left. v(x) = \Xi x \text{ if } \text{dist}(x, \partial(0, T)^d) < r\right\},$$

for any fixed $r > 0$, and

$$c_1(|\Xi|^p - 1) \leq h_{\text{hom}}^\delta(\Xi) \leq c_2(1 + |\Xi|^p)$$

with c_1, c_2 independent of δ, for $\delta \in [0, 1]$. Note that the independence of c_1 from δ is an immediate consequence of the Extension Theorem 6.3. Indeed, let $u_\varepsilon^\delta \to \Xi x$ be such that

$$h_{\text{hom}}^\delta(\Xi) = \lim_{\varepsilon \to 0} H_\varepsilon^\delta(u_\varepsilon^\delta, (0, 1)^d).$$

Applying Corollary 6.1 with $\Omega = (0, 1)^d$, we deduce that $T_\varepsilon u_\varepsilon^\delta$ converge to Ξx locally in $(0, 1)^d$ (in particular the convergence is strong e.g. in $(\frac{1}{4}, \frac{3}{4})^d$). Hence, using (6.44), the Extension Theorem (with r_0 in place of r), and the liminf inequality of the Γ-limit we have

$$\lim_{\varepsilon \to 0} H_\varepsilon^\delta(u_\varepsilon^\delta, (0, 1)^d)$$

$$\geq \liminf_{\varepsilon \to 0} H_\varepsilon(u_\varepsilon^\delta, (0, 1)^d)$$

$$\geq c_0 \liminf_{\varepsilon \to 0} \left(\frac{1}{\varepsilon^{p+d}} \int_{((0,1)^d \cap \varepsilon E)^2 \cap D_{r_0}} |u_\varepsilon^\delta(x) - u_\varepsilon^\delta(y)|^p dx dy - 1 \right)$$

$$\geq \frac{c_0}{c_2(r_0)} \liminf_{\varepsilon \to 0} \left(\frac{1}{\varepsilon^{p+d}} \int_{((\frac{1}{4}, \frac{3}{4})^d)^2 \cap D_R} |T_\varepsilon u_\varepsilon^\delta(x) - T_\varepsilon u_\varepsilon^\delta(y)|^p dx dy - 1 \right)$$

$$\geq \frac{c_0}{c_2(r_0)} \min \left\{ \frac{1}{2^d} c_R, 1 \right\} (|\Xi|^p - 1),$$

where in the last inequality we have used that

$$\Gamma\text{-}\lim_{\varepsilon \to 0} \frac{1}{\varepsilon^{p+d}} \int_{((\frac{1}{4}, \frac{3}{4})^d)^2 \cap D_R} |v(x) - v(y)|^p dx dy = c_R \int_{(\frac{1}{4}, \frac{3}{4})^d} |\nabla v|^p dx,$$

where c_R is as in (3.5).

Since h_{hom}^δ is increasing with δ, we may define

$$h_0(\Xi) = \inf_{\delta > 0} h_{\text{hom}}^\delta(\Xi) = \lim_{\delta \to 0^+} h_{\text{hom}}^\delta(\Xi),$$

and deduce (here we use the usual notation for the upper Γ-limit) that

$$\int_\Omega h_0(\nabla u)\, dx \geq \Gamma\text{-}\limsup_{\varepsilon \to 0} H_\varepsilon(u). \tag{6.50}$$

If $u \in W^{1,p}(\Omega; \mathbb{R}^m)$ and $u_\varepsilon \to u$ with $\sup_\varepsilon H_\varepsilon(u_\varepsilon) < +\infty$ then for all fixed Ω' compactly contained in Ω, if $R_0 < R$, upon identifying u_ε with its extension given by the Extension Theorem, we obtain that,

$$\int_{B_{R_0}} \int_{(\Omega')_\varepsilon(\xi)} \left| \frac{u_\varepsilon(x + \varepsilon \xi) - u_\varepsilon(x)}{\varepsilon} \right|^p dx\, d\xi \leq c,$$

so that

$$\liminf_{\varepsilon \to 0} H_\varepsilon(u_\varepsilon) \geq \liminf_{\varepsilon \to 0} H_\varepsilon(u_\varepsilon, \Omega') \geq \liminf_{\varepsilon \to 0} H_\varepsilon^\delta(u_\varepsilon, \Omega') - \delta c.$$

From this inequality we obtain (in terms of the lower Γ-limit)

$$\Gamma\text{-}\liminf_{\varepsilon \to 0} H_\varepsilon(u) \geq \int_\Omega h_0(\nabla u)dx$$

by the arbitrariness of δ and $\Omega' \Subset \Omega$. Hence, recalling (6.50), we have proved that

$$\Gamma\text{-}\lim_{\varepsilon \to 0} H_\varepsilon(u) = \int_\Omega h_0(\nabla u)dx,$$

and in particular that the Γ-limit exists as $\varepsilon \to 0$ (no subsequence is involved) and it can be represented as an integral functional with a homogeneous integrand. Note moreover that the lower-semicontinuity of the Γ-limit implies that h_0 is quasiconvex (see [4]).

We now prove that h_0 coincides with h_{hom} given by the asymptotic formula. First, note that

$$h_0(\Xi) \geq \limsup_{T \to +\infty} \frac{1}{T^d} \inf \left\{ \int_{(0,T)^d \cap E} \int_{(0,T)^d \cap E} h(x, y - x, v(y) - v(x))\, dx\, dy : \right.$$

$$\left. v(x) = \Xi x \text{ if } \mathrm{dist}(x, \partial(0, T)^d) < r \right\}. \tag{6.51}$$

If we take $r = k_0$, we obtain a lower bound for h_0.

To prove the opposite inequality, for any diverging sequence $\{T_j\}$ we can consider (almost-)minimizers v_j of the problems in (6.51) with $r = k_0$ and $T = T_j$. By Theorem 6.3 (applied component-wise and with $\varepsilon = 1$) with $\Omega = (0, T)^d$ and $\Omega' = (\frac{k_0}{2}, T_j - \frac{k_0}{2})^d$, we can consider the extended functions $\tilde{v}_j \in L^p((\frac{k_0}{2}, T_j - \frac{k_0}{2})^d; \mathbb{R}^m)$ with $\tilde{v}_j = v_j$ on $\Omega = (0, T)^d \cap E$ and

$$\int_{(\frac{k_0}{2}, T_j - \frac{k_0}{2})^d \cap D_R} |\tilde{v}_j(\xi) - \tilde{v}_j(\eta)|^p d\xi d\eta$$

$$\leq c_2(r_0) \int_{(0, T_j)^d \cap E)^2 \cap D_{r_0}} |v_j(\xi) - v_j(\eta)|^p d\xi d\eta \leq c\, T_j^d (1 + |\Xi|^p)$$

for some $c > 0$ independent of j. Upon choosing a larger $k_0 > 2$ we may suppose that $\lfloor \frac{k_0}{2} \rfloor + 1 < k_0$ so that we may consider $w_j \in L^p((0, T_j - n)^d; \mathbb{R}^m)$, where $n = 2\lfloor \frac{k_0}{2} \rfloor + 2$, defined by

$$w_j(x) = \tilde{v}_j\left(x + \left(\left\lfloor \frac{k_0}{2} \right\rfloor + 1\right)(1, \ldots, 1)\right) - \left(\left\lfloor \frac{k_0}{2} \right\rfloor + 1\right)\Xi(1, \ldots, 1).$$

Having set $\varepsilon_j = (T_j - n)^{-1}$ we can consider the scaled functions

$$u_j(x) = \varepsilon_j w_j\left(\frac{x}{\varepsilon_j}\right).$$

By the boundedness of the energies above and noting that there exists $c > 0$ such that $w_j(x) = \Xi x$ if $x \in E$ and $\mathrm{dist}(x, \partial(0, T_j - n)^d) < c$, upon extracting a subsequence, we may suppose that $u_j \to u$ and $u \in \Xi x + W_0^{1,p}((0, 1)^d; \mathbb{R}^m)$. We may then use the quasiconvexity inequality for h_0 to obtain

$$h_0(\Xi) \le \int_{(0,1)^d} h_0(\nabla u) dx$$

$$\le \liminf_j H_{\varepsilon_j}^\delta(u_j, (0, 1)^d)$$

$$\le \liminf_j H_{\varepsilon_j}(u_j, (0, 1)^d) + c\delta$$

$$\le \liminf_j \frac{1}{(T_j - n)^d} H_1(w_j, (0, T_j - n)^d) + c\delta$$

$$\le \liminf_j \frac{1}{(T_j - n)^d} H_1(v_j, (0, T_j)^d) + c\delta$$

$$= \liminf_j \frac{1}{(T_j - n)^d} \inf\left\{ \int_{(0,T_j)^d \cap E} \int_{(0,T_j)^d \cap E} h(x, y - x, v(y) - v(x)) \, dx \, dy \right.$$

$$\left. : v(x) = \Xi x \text{ if } \mathrm{dist}(x, \partial(0, T_j)^d) < k_0 \right\} + c\delta$$

$$= \liminf_j \frac{1}{T_j^d} \inf\left\{ \int_{(0,T_j)^d \cap E} \int_{(0,T_j)^d \cap E} h(x, y - x, v(y) - v(x)) \, dx \, dy \right.$$

$$\left. : v(x) = \Xi x \text{ if } \mathrm{dist}(x, \partial(0, T_j)^d) < k_0 \right\} + c\delta.$$

By the arbitrariness of δ and of the sequence T_j we obtain the desired upper bound for h_0, which, together with (6.51), proves the asymptotic formula.

In the convex case, again by Theorem 6.1, we may repeat the arguments used to get (6.51) to obtain the lower bound for h_0

$$h_0(\Xi) \geq \inf \left\{ \int_{(0,1)^d \cap E} \int_E h(x, y - x, v(y) - v(x)) \, dx \, dy : \right.$$
$$\left. v(x) - \Xi x \text{ is 1-periodic} \right\}. \tag{6.52}$$

Note that this implies that the right-hand side is bounded from above by $c_2(1+|\Xi|^p)$.

Now, let v be an (almost) minimizing function for (6.52), and set $v_\varepsilon(x) = \varepsilon v(\frac{x}{\varepsilon})$. After applying Theorem 6.3 to any set Ω compactly containing $(0, 1)^d$ to possibly redefine v_ε outside εE, we can suppose that v_ε converge in $L^p((0, 1)^d; \mathbb{R}^m)$ to Ξx and that

$$\frac{1}{\varepsilon^{p+d}} \int_{((0,1)^d \times (0,1)^d) \cap D_{\varepsilon R_0}} |v_\varepsilon(x) - v_\varepsilon(y)|^p \, dx \, dy \leq c(1 + |\Xi|^p).$$

We then estimate

$$h_{\text{hom}}^\delta(\Xi) \leq \liminf_{\varepsilon \to 0} H_\varepsilon^\delta(v_\varepsilon)$$
$$\leq \int_{(0,1)^d \cap E} \int_E h(x, y - x, v(y) - v(x)) \, dx \, dy + c\delta(1 + |\Xi|^p).$$

Taking the limit as $\delta \to 0$, we obtain the converse inequality of (6.52), and conclude the proof. $\qquad\square$

References

1. Acerbi, E., Chiadò Piat, V., Dal Maso, G., Percivale, D.: An extension theorem from connected sets, and homogenization in general periodic domains. Nonlinear Anal. **18**, 481–496 (1992)
2. Alicandro, R., Cicalese, M.: A general integral representation result for continuum limits of discrete energies with superlinear growth. SIAM J. Math. Anal. **36**, 1–37 (2004)
3. Alicandro, R., Focardi, M., Gelli, M.S.: Finite-difference approximation of energies in fracture mechanics. Ann. Sc. Norm. Super. Pisa Cl. Sci. (5) **29**, 671–709 (2000)
4. Braides, A., Defranceschi, A.: Homogenization of Multiple Integrals, volume 12 of Oxford Lecture Ser. Math. Appl. The Clarendon Press/Oxford University Press, New York (1998)
5. Braides, A., Gelli, M.S., Sigalotti, M.: The passage from nonconvex discrete systems to variational problem in Sobolev spaces: the one-dimensional case. Proc. Steklov Inst. Math. **236**, 395–414 (2002)
6. Braides, A., Chiadò Piat, V., D'Elia, L.: An extension theorem from connected sets and homogenization of non-local functionals. Nonlinear Anal. **208**, 112316 (2021)
7. Dal Maso, G.: An Introduction to Γ-convergence. Progr. Nonlinear Differential Equations Appl. Birkhäuser, Boston (1993)

Chapter 7
A Generalization and Applications to Point Clouds

Abstract In this chapter we prove a Γ-convergence result for functionals defined on point clouds. To this end, we first generalize the homogenization theorem for a family of convolution-type functionals where the Lebesgue measure is replaced by a measure with continuous density and we prove that it is Γ-asymptotically equivalent to a family of perturbed continuous functionals obtained by composing also the densities with suitable transportation maps. We then define discrete energies on point clouds whose number of points is going to infinity and prove that such a family Γ-converges to the same Γ-limit of the perturbed functionals.

Keywords Point clouds · Discrete functionals · Transportation maps · Voronoi cells · Γ-asymptotic equivalence

7.1 Perturbed Convolution-Type Functionals

We consider here a generalization of the class of functionals defined in (2.1), obtained by replacing the Lebesgue measure with a measure $\mu = \rho(x)\mathcal{L}^d$, with $\rho \in C^0(\Omega)$ and satisfying

$$0 < c \le \rho(x) \le C \quad \text{for every } x \in \Omega. \tag{7.1}$$

More precisely, given such a ρ, we set

$$F_\varepsilon[\rho](u) := \frac{1}{\varepsilon^d} \int_\Omega \int_\Omega f_\varepsilon\left(x, \frac{y-x}{\varepsilon}, \frac{u(y)-u(x)}{\varepsilon}\right)\rho(y)\rho(x)dx\,dy. \tag{7.2}$$

In the periodic case, that is when f_ε satisfies (6.1), a generalization of Theorem 6.1 is provided by the following result.

Theorem 7.1 *Let $F_\varepsilon[\rho]$ be defined by (7.2) with f_ε satisfying (6.1), $\rho \in C^0(\Omega)$ and such that (7.1) holds. Then, under the assumptions of Theorem 6.1,*

$$\Gamma(L^p)\text{-}\lim_{\varepsilon \to 0} F_\varepsilon[\rho](u) = \begin{cases} \int_\Omega f_{\text{hom}}(\nabla u(x))\rho^2(x)dx & \text{if } u \in W^{1,p}(\Omega; \mathbb{R}^m), \\ +\infty & \text{otherwise}, \end{cases}$$

where f_{hom} is defined by (6.7).

Proof We highlight only the main differences with respect to the proof of Theorem 6.1. Note that, by (7.1), $F_\varepsilon[\rho]$ satisfies all the assumptions of Theorem 5.1. Hence, given $\varepsilon_j \to 0$, there exists a subsequence (not relabelled) such that

$$\Gamma(L^p)\text{-}\lim_{j \to +\infty} F_{\varepsilon_j}[\rho](u, A) = \int_A f_0(x, \nabla u(x))\, dx.$$

The characterization of non-homogeneous quasiconvex functions by their minima (see [2, Theorem II]) yields that for every $M \in \mathbb{R}^{m \times d}$ and for almost every $x_0 \in \Omega$

$$f_0(x_0, M) = \lim_{r \to 0} \frac{1}{r^d} \min \left\{ \int_{Q_r(x_0)} f_0(x, \nabla u(x))dx : u - Mx \in W_0^{1,p}(Q_r(x_0); \mathbb{R}^m) \right\}.$$

Then, proceeding as in the proof of Theorem 6.1 and using the continuity of ρ, we obtain that

$$f_0(x_0, M) = \rho^2(x_0) f_{\text{hom}}(M).$$

Since f_0 does not depend on $(\varepsilon_j)_j$, we get the conclusion. $\qquad\square$

We study now a perturbed version of the class of functionals defined in (7.2), obtained by composing the energy densities with a family of transportation maps. Specifically, we consider the family of perturbed functionals $E_\varepsilon : L^p(\Omega; \mathbb{R}^m) \to [0, +\infty]$ defined by

$$E_\varepsilon(u) := \frac{1}{\varepsilon^d} \int_\Omega \int_\Omega f_\varepsilon\left(x, \frac{T_\varepsilon(y) - T_\varepsilon(x)}{\varepsilon}, \frac{u(y) - u(x)}{\varepsilon}\right)\rho(y)\rho(x)dx\, dy. \tag{7.3}$$

Note that in Sect. 6.4 we denote by T_ε the extension operator $T_\varepsilon : L^p(\Omega \cap \varepsilon E) \to L^p(\Omega)$. Here $T_\varepsilon : \Omega \to \Omega$ is a measurable map (representing a transportation map in Sect. 7.2). We also assume that f_ε fulfills assumptions (H0)–(H2), with $\psi_{\varepsilon,2}$ satisfying the additional hypothesis

$$\psi_{\varepsilon,2}(\xi) := \overline{\psi}_\varepsilon(|\xi|), \text{ where } \overline{\psi}_\varepsilon : [0, +\infty) \to [0, +\infty) \text{ is non increasing} \tag{7.4}$$

and $\overline{\psi}_\varepsilon(0) < c$, for some constant $c > 0$. Notice that, under these additional assumptions we have that $\lim \sup_{\varepsilon \to 0} \int_{\mathbb{R}^d} \psi_{\varepsilon,2}(\xi) \, d\xi < +\infty$.

Remark 7.1 Set

$$g_\varepsilon(x, \xi, z) = f_\varepsilon\left(x, \frac{T_\varepsilon(x + \varepsilon\xi) - T_\varepsilon(x)}{\varepsilon}, z\right)\rho(x)\rho(x + \varepsilon\xi) \tag{7.5}$$

and note that, under the assumptions above on f_ε and ρ, g_ε satisfies (H0)–(H2) provided the following regularity conditions are fulfilled by T_ε:

$$|T_\varepsilon(y) - T_\varepsilon(x)| \leq C'(|y - x| + \varepsilon), \quad \text{if } |y - x| \leq \varepsilon r' \tag{7.6}$$

$$|T_\varepsilon(y) - T_\varepsilon(x)| \geq C''|y - x|, \quad \text{if } |y - x| \geq \varepsilon r'', \tag{7.7}$$

for some C', C'', r', r'' positive constants. Indeed, if r'_0 is a positive constant such that $r'_0 < r'$ and $C'(r'_0 + 1) < r_0$, then by (7.6)

$$\left|\frac{T_\varepsilon(x + \varepsilon\xi) - T_\varepsilon(x)}{\varepsilon}\right| \leq r_0 \text{ for every } \xi \in B_{r'_0}$$

and assumption (H0) on f_ε yields that the same assumption is satisfied by g_ε with another choice of the constants. Moreover, denoting

$$\tilde{\psi}_\varepsilon(\xi) := \sup_{x \in \Omega} \overline{\psi}_\varepsilon\left(\left|\frac{T_\varepsilon(x + \varepsilon\xi) - T_\varepsilon(x)}{\varepsilon}\right|\right), \tag{7.8}$$

from conditions (7.7) and the monotonicity of $\overline{\psi}_\varepsilon$ we get

$$\int_{\mathbb{R}^d} \tilde{\psi}_\varepsilon(\xi)|\xi|^p \, d\xi \leq \int_{B_{r''}} \overline{\psi}_\varepsilon(0)|\xi|^p \, d\xi + \int_{B_{r''}^c} \overline{\psi}_\varepsilon(C''|\xi|)|\xi|^p \, d\xi$$

which yields that condition (2.5) is satisfied by $\tilde{\psi}_\varepsilon$, since it is satisfied by $\psi_{\varepsilon,2}$. Analogously it can be shown that $\tilde{\psi}_\varepsilon$ satisfies (H2). Hence, under the assumptions (7.6) and (7.7), energies as in (7.3) belong to the class of functionals satisfying the hypotheses of Theorem 5.1. Notice that conditions (7.6) and (7.7) hold in particular if $\|T_\varepsilon - id\|_\infty \leq C\varepsilon$.

In the next one-dimensional example we show that the asymptotic behaviour of E_ε could be degenerate if (7.4) is not satisfied.

Example 7.1 (A Pathological Problem) Assume that in (7.3) $\Omega = (0, 1)$, $\rho \equiv 1$ and $f_\varepsilon(x, \xi, z) = a(\xi)|z|^p$ with $a : \mathbb{R} \to [0, +\infty)$ defined as follows

$$a(\xi) = \begin{cases} 1 & \xi \in (-1, 1) \cup \mathbb{Q} \\ 0 & \text{otherwise.} \end{cases}$$

Given $\lambda_\varepsilon \to 0$, let $T_\varepsilon : (0, 1) \to (0, 1)$ be such that $T_\varepsilon((0, 1)) \subset \varepsilon\mathbb{Q}$ and $\|T_\varepsilon - id\|_\infty \leq \lambda_\varepsilon$. In particular (7.6) and (7.7) are satisfied if $\lambda_\varepsilon = o(\varepsilon)$. We may construct such maps as follows: for any $k \in \{0, \ldots, \lceil \lambda_\varepsilon^{-1} \rceil - 1\}$, let $q_{k,\varepsilon} \in \mathbb{Q} \cap \left(\varepsilon^{-1}\lambda_\varepsilon[k, k + 1)\right) \cap (0, \varepsilon^{-1})$ and set

$$T_\varepsilon(x) := \varepsilon q_{k,\varepsilon} \text{ if } x \in \lambda_\varepsilon[k, k + 1) \text{ for some } k \in \{0, \ldots, \lceil \lambda_\varepsilon^{-1} \rceil - 1\}.$$

Since $a = \chi_{(-1,1)}$ almost everywhere, $G_\varepsilon[a] = G_\varepsilon^1$; thus, by Theorem 3.1, $\Gamma\text{-}\lim_{\varepsilon \to 0} G_\varepsilon[a](u) < +\infty$ for any $u \in W^{1,p}(0, 1)$. Whereas, since $(T_\varepsilon y - T_\varepsilon x)/\varepsilon \in \mathbb{Q}$, E_ε reads

$$E_\varepsilon(u) = \int_{-\infty}^{+\infty} \int_{0 \vee (-\varepsilon\xi)}^{1 \wedge (1 - \varepsilon\xi)} \left| \frac{u(x + \varepsilon\xi) - u(x)}{\varepsilon} \right|^p dx \, d\xi \, ,$$

From which we deduce that

$$\Gamma\text{-}\lim_{\varepsilon \to 0} E_\varepsilon(u) = \begin{cases} 0 & \text{if } u' = 0 \text{ in } (0, 1), \\ +\infty & \text{otherwise.} \end{cases}$$

In the next proposition we show that if $\|T_\varepsilon - id\|_\infty = o(\varepsilon)$ and $f_\varepsilon(x, \cdot, z)$ satisfies a suitable continuity assumption uniformly with respect to ε and x, then the functionals E_ε defined by (7.3) are asymptotically equivalent in the sense of the Γ-convergence to the functionals $F_\varepsilon[\rho]$ defined by (7.2).

Proposition 7.1 *Let $F_\varepsilon[\rho]$ and E_ε be defined by (7.2) and (7.3), respectively, with f_ε satisfying (H0)–(H2) and (7.4). Assume in addition that:*

(i) *there exists a family of positive functions $\{\omega_h\}_{h>0} \subset L^1_{loc}(\mathbb{R}^d)$ such that $\omega_h \to 0$ in $L^1_{loc}(\mathbb{R}^d)$ and*

$$\sup_{x \in \Omega} \sup_{|v| \leq h} |f_\varepsilon(x, \xi + v, z) - f_\varepsilon(x, \xi, z)| \leq \omega_h(\xi)|z|^p \tag{7.9}$$

for every $\varepsilon > 0$ and $z \in \mathbb{R}^m$;

(ii) $\left\| \dfrac{T_\varepsilon - id}{\varepsilon} \right\|_{L^\infty(\Omega;\mathbb{R}^d)} \to 0.$

Then

$$\Gamma(L^p)\text{-}\liminf_{\varepsilon \to 0} E_\varepsilon(u) = \Gamma(L^p)\text{-}\liminf_{\varepsilon \to 0} F_\varepsilon[\rho](u),$$

$$\Gamma(L^p)\text{-}\limsup_{\varepsilon \to 0} E_\varepsilon(u) = \Gamma(L^p)\text{-}\limsup_{\varepsilon \to 0} F_\varepsilon[\rho](u).$$

Proof Notice that, since both $F_\varepsilon[\rho]$ and E_ε satisfy the hypotheses of Proposition 3.3, we have that

$$\Gamma(L^p)\text{-}\lim_{\varepsilon\to 0} E_\varepsilon(u) = \Gamma(L^p)\text{-}\lim_{\varepsilon\to 0} F_\varepsilon[\rho](u) = +\infty, \ u \in L^p(\Omega; \mathbb{R}^m)\backslash W^{1,p}(\Omega; \mathbb{R}^m).$$

Hence, by (H0)–(H1), it suffices to prove that

$$E_\varepsilon(u_\varepsilon) = F_\varepsilon[\rho](u_\varepsilon) + o(1) \tag{7.10}$$

for any sequence $(u_\varepsilon)_\varepsilon$ such that $G_\varepsilon^r(u_\varepsilon)$ is uniformly bounded for some $r \leq r_0 \wedge r_0'$, where r_0 and r_0' refer to assumption (H0) for $F_\varepsilon[\rho]$ and E_ε, respectively.

By Lemma 5.1, we may reduce to prove (7.10) in the case $f_\varepsilon(x, \xi, z) = 0$ for every $x \in \Omega$, $|\xi| > T$ and $z \in \mathbb{R}^m$, for some $T > 0$. Let then u_ε be such that $\sup_{\varepsilon>0} G_\varepsilon^r(u_\varepsilon) < +\infty$. We have

$$E_\varepsilon(u_\varepsilon) = F_\varepsilon(u_\varepsilon) + R_\varepsilon(u_\varepsilon),$$

where

$$R_\varepsilon(v) := \int_{B_T} \int_{\Omega_\varepsilon(\xi)} \left(g_\varepsilon\left(x, \xi, \frac{v(x+\varepsilon\xi) - v(x)}{\varepsilon}\right) \right.$$

$$\left. - f_\varepsilon\left(x, \xi, \frac{v(x+\varepsilon\xi) - v(x)}{\varepsilon}\right) \right) dx \, d\xi ,$$

with g_ε defined by (7.5). Set

$$h_\varepsilon := \left\| \frac{T_\varepsilon - id}{\varepsilon} \right\|_{L^\infty(\Omega; \mathbb{R}^d)}.$$

By (i) and Lemma 4.1 we get

$$|R_\varepsilon(u_\varepsilon)| \leq C \int_{B_T} \int_{\Omega_\varepsilon(\xi)} \omega_{2h_\varepsilon}(\xi) \left| \frac{u_\varepsilon(x+\varepsilon\xi) - u_\varepsilon(x)}{\varepsilon} \right|^p dx \, d\xi$$

$$\leq C(R^p + 1) G_\varepsilon^r(u_\varepsilon) \int_{B_T} \omega_{2h_\varepsilon}(\xi) d\xi,$$

which goes to zero as $\varepsilon \to 0$ by (ii). □

As a straightforward consequence of the previous proposition and Theorem 7.1, we obtain the following result.

Corollary 7.1 *Under the assumptions of Proposition 7.1, assume in addition that f_ε satisfies (6.1). Then*

$$\Gamma(L^p)\text{-}\lim_{\varepsilon \to 0} E_\varepsilon(u) = \begin{cases} \displaystyle\int_\Omega f_{\text{hom}}(\nabla u(x))\rho^2(x)dx & \text{if } u \in W^{1,p}(\Omega; \mathbb{R}^m), \\ +\infty & \text{otherwise,} \end{cases}$$

where f_{hom} is defined by (6.7).

7.2　Application to Functionals Defined on Point Clouds

The case in which $T_\varepsilon(\Omega) = X_{n_\varepsilon} := \{x_i\}_{i=1}^{n_\varepsilon} \subset \Omega$, has already been studied in the context of variational methods for Machine Learning, when dealing with discrete convolution-type energies of the form

$$\frac{1}{\varepsilon^p} \frac{1}{n_\varepsilon^2} \sum_{i,j=1}^{n_\varepsilon} a_{i,j}^\varepsilon |u(x_i) - u(x_j)|^p, \tag{7.11}$$

that are the discrete version of energies (7.3) when T_ε are transportation maps from Ω to X_{n_ε}, see [4] when $p = 1$ and [1] when $p > 1$. Therein, X_{n_ε} denotes a point cloud obtained by refining random samples of a given probability measure $\mu \ll \mathcal{L}^d$, having continuous density bounded from above and below by two positive constants.

In this subsection, we will prove a Γ-convergence result for a generalized version of discrete energies as in (7.11) defined on point clouds. In particular, we will recover the convergence result provided in [1]. Before setting the problem, we recall, for the reader's convenience, some useful notions about point-cloud models.

Let $\mu = \rho(x)\mathcal{L}^d$ be a probability measure supported on Ω, such that $\rho \in C^0(\Omega)$ and satisfies (7.1). Given $(X, \sigma, \mathbb{P})$ a probability space, we consider a sequence of random variables

$$x_i : X \ni \omega \mapsto x_i(\omega) \in \Omega, \quad i \in \mathbb{N},$$

that are *i.i.d.* according to the distribution μ. Then, given $n \in \mathbb{N}$, we say that the set $X_n(\omega) = \{x_i(\omega)\}_{i=1}^n$ is a *point cloud* obtained as samples from a given distribution μ. In the following, we will drop the dependence on ω for the sake of simplicity of notation, unless otherwise specified. To any point cloud X_n we associate its *empirical measure*

$$\mu_n = \frac{1}{n} \sum_{i=1}^n \delta_{x_i}. \tag{7.12}$$

It is well known that μ_n weak* converge to μ as $n \to +\infty$ $\mathbb{P}$-almost surely.

Let $n_\varepsilon \in \mathbb{N}$ be such that

$$\lim_{\varepsilon \to 0} n_\varepsilon = +\infty$$

and let $f : \mathbb{R}^d \times \mathbb{R}^m \to [0, +\infty)$ be a Borel function that is convex in the second variable. We then consider the family of functionals ("dis" stands for discrete)

$$E_\varepsilon^{\mathrm{dis}}(u) = \frac{1}{\varepsilon^d n_\varepsilon^2} \sum_{i,j=1}^{n_\varepsilon} f\left(\frac{x_i - x_j}{\varepsilon}, \frac{u(x_i) - u(x_j)}{\varepsilon}\right),$$

defined on functions $u : X_{n_\varepsilon} \to \mathbb{R}^m$. Note that $E_\varepsilon^{\mathrm{dis}}$ can be written in terms of μ_{n_ε} as

$$E_\varepsilon^{\mathrm{dis}}(u) = \frac{1}{\varepsilon^d} \int_\Omega \int_\Omega f\left(\frac{y - x}{\varepsilon}, \frac{u(y) - u(x)}{\varepsilon}\right) d\mu_{n_\varepsilon}(x)\, d\mu_{n_\varepsilon}(y). \qquad (7.13)$$

Let E_ε be defined by (7.3) with

$$f_\varepsilon(x, \xi, z) = f(\xi, z). \qquad (7.14)$$

If $T_\varepsilon : \Omega \to X_{n_\varepsilon}$ is a transportation map between μ_{n_ε} and μ; that is, $(T_\varepsilon)_{\#}\mu = \mu_{n_\varepsilon}$, where $T_{\#}\mu$ denotes the push-forward of μ by T, then we may identify any $u : X_{n_\varepsilon} \to \mathbb{R}^m$ with its piecewise constant interpolation on $T_\varepsilon^{-1}(x_i)$, $i = 1, \ldots, n_\varepsilon$, and, by (7.13),

$$E_\varepsilon^{\mathrm{dis}}(u) = E_\varepsilon(u) \quad \text{for every } u \in PC(X_{n_\varepsilon}), \qquad (7.15)$$

where

$$PC(X_{n_\varepsilon}) := \left\{u : \Omega \to \mathbb{R}^m \mid u \text{ is constant on } T_\varepsilon^{-1}(x_i) \text{ for every } 1 \le i \le n_\varepsilon\right\}.$$

With a slight abuse of notation we assume that $E_\varepsilon^{\mathrm{dis}}$ is defined on the whole space $L^p(\Omega; \mathbb{R}^m)$ by setting

$$E_\varepsilon^{\mathrm{dis}}(u) = \begin{cases} \dfrac{1}{\varepsilon^d n_\varepsilon^2} \displaystyle\sum_{i,j=1}^{n_\varepsilon} f\left(\dfrac{x_i - x_j}{\varepsilon}, \dfrac{u(x_i) - u(x_j)}{\varepsilon}\right), & u \in PC(X_{n_\varepsilon}) \\ +\infty & \text{otherwise.} \end{cases} \qquad (7.16)$$

In the next theorem we show that $E_\varepsilon^{\mathrm{dis}}$ and E_ε are asymptotically equivalent in the sense of Γ-convergence under the assumptions of Proposition 7.1.

Theorem 7.2 *Let X_{n_ε} be a family of point clouds obtained as samples from μ and let $T_\varepsilon : \Omega \to X_{n_\varepsilon}$ be a transportation map between μ_{n_ε} and μ, where μ_{n_ε} is* *defined in (7.12). Let $E_\varepsilon^{\mathrm{dis}}$ be defined by (7.16) and let E_ε be defined by (7.3) with f_ε satisfying (7.14) and $f(\xi, \cdot)$ convex for any $\xi \in \mathbb{R}^d$. If f_ε and T_ε satisfy the assumptions of Proposition 7.1, then*

$$\Gamma(L^p)\text{-}\lim_{\varepsilon \to 0} E_\varepsilon^{\mathrm{dis}}(u) = \Gamma(L^p)\text{-}\lim_{\varepsilon \to 0} E_\varepsilon(u)$$

$$= \begin{cases} \displaystyle\int_{\mathbb{R}^d} \int_\Omega f(\xi, \nabla u(x)\xi)\rho^2(x)\, dx\, d\xi & \text{if } u \in W^{1,p}(\Omega; \mathbb{R}^m), \\ +\infty & \text{otherwise.} \end{cases}$$
$$(7.17)$$

Proof The second equality in (7.17) is a straightforward consequence of Corollary 7.1, Theorem 6.2 and Remark 6.1. Since $E_\varepsilon^{\mathrm{dis}} \geq E_\varepsilon$, in order to prove the first equality in (7.17) it suffices to prove that given $u \in W^{1,p}(\Omega)$ we can find $(u_\varepsilon)_\varepsilon \subset PC(X_{n_\varepsilon})$ such that $u_\varepsilon \to u$ in $L^p(\Omega; \mathbb{R}^m)$ and

$$\limsup_{\varepsilon \to 0} E_\varepsilon^{\mathrm{dis}}(u_\varepsilon) \leq \int_{\mathbb{R}^d} \int_\Omega f(\xi, \nabla u(x)\xi)\rho^2(x)\, dx\, d\xi. \tag{7.18}$$

By a density argument it suffices to prove (7.18) for $u \in C^\infty(\mathbb{R}^d; \mathbb{R}^m)$. Fix such a function u and note that, by the regularity assumptions on f and assumption (ii) of Proposition 7.1, we have that

$$\lim_{\varepsilon \to 0} E_\varepsilon(u) = \int_{\mathbb{R}^d} \int_\Omega f(\xi, \nabla u(x)\xi)\rho^2(x)\, dx\, d\xi. \tag{7.19}$$

Let $u_\varepsilon \in PC(X_{n_\varepsilon})$ be defined by

$$u_\varepsilon(x_i) := \frac{1}{|V_\varepsilon^i|} \int_{V_\varepsilon^i} u(y)\, dy \quad i \in \{1, \dots, n_\varepsilon\},$$

where, for $i = 1, \dots, n_\varepsilon$, we have set $V_\varepsilon^i = T_\varepsilon^{-1}(x_i)$.

Note that $V_\varepsilon^i \subseteq B_{r_\varepsilon}(x_i)$, where $r_\varepsilon := \|T_\varepsilon - Id\|_\infty$. Hence, $\|u_\varepsilon - u\|_{L^\infty(\Omega; \mathbb{R}^m)} \leq Cr_\varepsilon \to 0$ by assumption (ii) of Proposition 7.1 By the convexity of $f(\xi, \cdot)$, we get

$$f\left(\frac{x_i - x_j}{\varepsilon}, \frac{u_\varepsilon(x_i) - u_\varepsilon(x_j)}{\varepsilon}\right)$$

$$= f\left(\frac{x_i - x_j}{\varepsilon}, \frac{1}{|V_\varepsilon^i|}\int_{V_\varepsilon^i} \frac{u(y)}{\varepsilon}\, dy - \frac{1}{|V_\varepsilon^j|}\int_{V_\varepsilon^j} \frac{u(x)}{\varepsilon}\, dx\right)$$

$$\leq \frac{1}{|V_\varepsilon^i|}\int_{V_\varepsilon^i} f\left(\frac{x_i - x_j}{\varepsilon}, \frac{u(y)}{\varepsilon} - \frac{1}{|V_\varepsilon^j|}\int_{V_\varepsilon^j} \frac{u(x)}{\varepsilon}\, dx\right) dy \tag{7.20}$$

$$\leq \frac{1}{|V_\varepsilon^i||V_\varepsilon^j|}\int_{V_\varepsilon^i}\int_{V_\varepsilon^j} f\left(\frac{x_i - x_j}{\varepsilon}, \frac{u(y) - u(x)}{\varepsilon}\right) dy\, dx$$

for any $1 \leq i, j \leq n_\varepsilon$.

Note that, since T_ε is a transportation map between μ and μ_ε, by the continuity of ρ we get

$$\frac{1}{|V_\varepsilon^i|} = (\rho(x_i) + o(1)) \, n_\varepsilon,$$

uniformly in $1 \le i \le n_\varepsilon$. Hence, by (7.20) we get

$$E_\varepsilon^{\text{dis}}(u_\varepsilon) = \frac{1}{\varepsilon^d n_\varepsilon^2} \sum_{i,j=1}^{n_\varepsilon} f\left(\frac{x_i - x_j}{\varepsilon}, \frac{u_\varepsilon(x_i) - u_\varepsilon(x_j)}{\varepsilon}\right)$$

$$\le \frac{1}{\varepsilon^d n_\varepsilon^2} \sum_{i,j=1}^{n_\varepsilon} \frac{1}{|V_\varepsilon^i||V_\varepsilon^j|} \int_{V_\varepsilon^i} \int_{V_\varepsilon^j} f\left(\frac{x_i - x_j}{\varepsilon}, \frac{u(y) - u(x)}{\varepsilon}\right) dy \, dx$$

$$= \frac{1}{\varepsilon^d} \sum_{i,j=1}^{n_\varepsilon} \int_{V_\varepsilon^i} \int_{V_\varepsilon^j} f\left(\frac{x_i - x_j}{\varepsilon}, \frac{u(y) - u(x)}{\varepsilon}\right) \rho(x_i) \rho(x_j) \, dy \, dx + o(1)$$

$$= E_\varepsilon(u_\varepsilon) + o(1),$$

and the conclusion follows from (7.19). $\qquad\qquad\square$

Remark 7.2 In [3], extending previous results, it has been proved that for $d \ge 2$ almost surely there exists a family of transportation maps $T_n(\omega) : \Omega \to X_n(\omega)$ between μ and μ_n such that

$$0 < c \le \liminf_{n \to +\infty} \frac{\|T_n(\omega) - id\|_{L^\infty}}{l_n} \le \limsup_{n \to +\infty} \frac{\|T_n(\omega) - id\|_{L^\infty}}{l_n} \le C,$$

where

$$l_n = \begin{cases} \dfrac{(\log n)^{3/4}}{n^{1/2}} & \text{if } d = 2, \\[2ex] \left(\dfrac{\log n}{n}\right)^{1/d} & \text{if } d \ge 3. \end{cases}$$

Hence, if $n_\varepsilon \to +\infty$ and

$$\lim_{\varepsilon \to 0} \frac{l_{n_\varepsilon}}{\varepsilon} = 0, \tag{7.21}$$

the corresponding transportation maps $T_\varepsilon = T_{n_\varepsilon}$ are such that $\|T_\varepsilon - id\|_{L^\infty} = o(\varepsilon)$ and, in particular, assumptions (ii) of Proposition 7.1 is satisfied. Hence if (7.21) is satisfied the Γ-convergence result stated in Theorem 7.2 holds almost surely, thus extending the convergence result provided in [1], where the analysis is limited to energy densities of the form $f(\xi, z) = a(|\xi|)|z|^p$, with $a(\cdot)$ non increasing.

References

1. Crook, O.M., Hurst, T., Schönlieb, C.-B., Thorpe, M., Zygalakis, K.C.: Pde-inspired algorithms for semi-supervised learning on point clouds (2019). Preprint, arXiv:1909.10221v1
2. Dal Maso, G., Modica, L.: Integral functionals determined by their minima. Rend. Semin. Mat. Univ. Padova **76**, 255–267 (1986)
3. García Trillos, N., Slepčev, D.: On the rate of convergence of empirical measures in ∞-transportation distance. Canad. J. Math. **67**, 1358–1383 (2015)
4. García Trillos, N., Slepčev, D.: Continuum limit of total variation on point clouds. Arch. Ration. Mech. Anal. **220**, 193–241 (2016)

Chapter 8
Stochastic Homogenization

Abstract In this section we consider random energies of convolution type and prove that, under stationarity and ergodicity assumptions, the Γ-limit of such energies is almost surely a deterministic integral functional whose integrand can be characterized through an asymptotic formula.

Keywords Stochastic homogenization · Statistically homogeneous random functions · Ergodicity

Let $(X, \sigma, \mathbb{P})$ be a standard probability space with a measure-preserving ergodic dynamical system τ_y, $y \in \mathbb{R}^d$. We recall that $\{\tau_y\}$ is a collection of measurable invertible maps $\tau_y : X \mapsto X$ such that

- $\tau_{y+y'} = \tau_y \circ \tau_{y'}$ for all $y, y' \in \mathbb{R}^d$, $\tau_0 = \text{Id}$,
- $\mathbb{P}(\tau_y(A)) = \mathbb{P}(A)$ for all $A \in \sigma$ and $y \in \mathbb{R}^d$,
- $\tau : X \times \mathbb{R}^d \mapsto X$ is measurable, where $\mathbb{R}^d$ is equipped with the Borel σ-algebra.

The ergodicity of τ means that for every $A \in \sigma$ such that $\tau_y(A) = A$ for all $y \in \mathbb{R}^d$ there holds either $\mathbb{P}(A) = 0$ or $\mathbb{P}(A) = 1$.

Consider now a random function f defined as a measurable map

$$f : X \times \mathbb{R}^d \times \mathbb{R}^m \to [0, +\infty),$$

where $\mathbb{R}^d$ and $\mathbb{R}^m$ are equipped with the Borel σ-algebra. We assume that there exist two positive constants r_0 and c_0 and four random functions $\rho_1(\omega), \psi_1(\omega) : B_{r_0} \to [0, +\infty)$ and $\psi_2(\omega), \rho_2(\omega) : \mathbb{R}^d \to [0, +\infty)$ such that the following conditions hold $\mathbb{P}$-almost surely:

$$\psi_1(\omega)(\xi)|z|^p - \rho_1(\omega)(\xi) \leq f(\omega, \xi, z), \quad \text{for a.e. } \xi \in B_{r_0}, \tag{8.1}$$

$$f(\omega, \xi, z) \leq \psi_2(\omega)(\xi)|z|^p + \rho_2(\omega)(\xi), \quad \text{for a.e. } \xi \in \mathbb{R}^d, \tag{8.2}$$

$$\rho_1 \in L^1(X \times B_{r_0}) \quad \text{and} \quad \psi_1(\omega)(\xi) \geq c_0, \quad \text{for a.e. } \xi \in B_{r_0}, \tag{8.3}$$

$$C_1(\cdot) := \int_{\mathbb{R}^d} \psi_2(\cdot)(\xi)|\xi|^p + \rho_2(\cdot)(\xi)\, d\xi \in L^1(X, \mathbb{P}), \tag{8.4}$$

requesting additionally that

$$\text{if } \int_{B_{r_0}} \psi_2(\omega)(\xi)\, d\xi = +\infty \text{ then there exists } c_1 > 0 \text{ such that}$$
$$\psi_2(\omega)(\xi) \leq c_1 \psi_1(\omega)(\xi) \text{ for a.e. } \xi \in B_{r_0}. \tag{8.5}$$

Letting $f(\omega)(y, \xi, z) = \mathfrak{f}(\tau_y\omega, \xi, z)$, we introduce a family of stochastic non-local energy functionals $F_\varepsilon(\omega) : L^p_{\text{loc}}(\mathbb{R}^d; \mathbb{R}^m) \times \mathcal{A}(\mathbb{R}^d) \to [0, +\infty), \varepsilon > 0$, defined by

$$F_\varepsilon(\omega)(u, A) := \int_{\mathbb{R}^d} \int_{A_\varepsilon(\xi)} f(\omega)\left(\frac{x}{\varepsilon}, \xi, \frac{u(x + \varepsilon\xi) - u(x)}{\varepsilon}\right) dx\, d\xi. \tag{8.6}$$

By construction the densities f are *statistically homogeneous* random functions of x that is

$$f(\omega)(x + y, \xi, z) = f(\tau_y\omega)(x, \xi, z), \quad \text{for every } \omega \in X, \, x, y, \xi \in \mathbb{R}^d, \, z \in \mathbb{R}^m. \tag{8.7}$$

In Theorem 8.2 below we prove a homogenization theorem for the functionals F_ε. Our proof of a homogenization formula relies on a subadditive ergodic theorem, a result that we recall below preceded by the definition of multiparameter stationary stochastic processes.

Definition 8.1 (Subadditive Process) Let $\mathcal{V}$ be the family of all finite subset of the lattice $\mathbb{Z}^{d,+} := \{0, 1, \dots\}^d$. A real valued process $\Psi : \mathcal{V} \to L^1(X, \mathbb{P})$ is called a *subadditive process* if it satisfies the following conditions:

(i) it is stationary, that is for any $j \in \mathbb{Z}^{d,+}$ and any finite collection $\{V_1, \dots, V_N\} \subset \mathcal{V}$ the joint law of $\{\Psi(V_1 + j), \dots, \Psi(V_N + j)\}$ is the same as the joint law of $\{\Psi(V_1), \dots, \Psi(V_N)\}$;
(ii) it is subadditive, that is $\Psi(V_1 \cup V_2) \leq \Psi(V_1) + \Psi(V_2)$ for any disjoint V_1 and V_2 in $\mathcal{V}$;
(iii) there holds

$$\inf_{n \in \mathbb{N}} \int_X \frac{1}{n^d} \Psi(\{0, 1, \dots n\}^d)(\omega)\, d\mathbb{P}(\omega) > -\infty.$$

A proof of the following theorem can be found in [3, Theorem 1].

Theorem 8.1 (Subadditive Ergodic Theorem) *Let $\mathcal{B}_0$ be a family of Borel subsets of $[0, 1]^d$ such that*

$$\sup\{|\partial B + B_\delta| \ : \ B \in \mathcal{B}_0\} \to 0, \quad as \ \delta \to 0.$$

Then, for every subadditive process $\Psi : \mathcal{V} \to L^1(X, \mathbb{P})$ there exists a real random variable $\phi \in L^1(X, \mathbb{P})$ such that

$$\sup\left\{\left|\frac{1}{N^d}\Psi((NB) \cap \mathbb{Z}^{d,+}) - |B|\phi\right| : B \in \mathcal{B}_0\right\} \to 0 \quad almost \ surely \ as \ N \to +\infty.$$
$$(8.8)$$

The stochastic homogenization theorem is then stated as follows. This result extends [1, Theorem 6.1], where quadratic convolution energies with random coefficients are studied.

Theorem 8.2 (Stochastic Homogenization Theorem) *Let f be a random function satisfying (8.1)–(8.5). Let F_ε be defined as in (8.6) with $f(\omega)(y, \cdot, \cdot) = f(\tau_y\omega, \cdot, \cdot)$ a statistically homogeneous random function according to (8.7). Then for $\mathbb{P}$-almost every $\omega \in X$ and for every $M \in \mathbb{R}^{m\times d}$ the limit*

$$f_{\mathrm{hom}}(\omega)(M) = \lim_{R \to +\infty} \frac{1}{R^d} \inf\left\{F_1(\omega)(u, Q_R) : u \in \mathcal{D}^{1,M}(Q_R)\right\},\qquad (8.9)$$

where $\mathcal{D}^{1,M}(Q_R)$ is defined by (5.19), exists and defines a quasiconvex function $f_{\mathrm{hom}}(\omega) : \mathbb{R}^{m\times d} \to [0, +\infty)$ satisfying

$$c(\omega)(|M|^p - 1) \leq f_{\mathrm{hom}}(M) \leq C(\omega)(|M|^p + 1), \qquad (8.10)$$

where $c(\cdot), C(\cdot) \in L^1(X, \mathbb{P})$. Moreover, for $\mathbb{P}$-almost every $\omega \in X$ and for every $A \in \mathcal{A}^{\mathrm{reg}}(\Omega)$ there holds

$$\Gamma\text{-}\lim_{\varepsilon \to 0} F_\varepsilon(\omega)(u, A) = F(\omega)(u, A) := \begin{cases} \int_A f_{\mathrm{hom}}(\omega)(\nabla u(x))dx & u \in W^{1,p}(A; \mathbb{R}^m) \\ +\infty & otherwise. \end{cases}$$

If, in addition, the dynamical system τ is ergodic, then $f_{\mathrm{hom}}(\omega)(\cdot)$ is constant almost surely and satisfies

$$f_{\mathrm{hom}}(\omega)(M) \equiv f_{\mathrm{hom}}(M)$$

$$:= \lim_{R \to +\infty} \frac{1}{R^d} \int_X \inf\left\{F_1(\omega)(u, Q_R) : u \in \mathcal{D}^{1,M}(Q_R)\right\}d\mathbb{P}(\omega).$$
$$(8.11)$$

Proof For any given $T > 0$ let $F_\varepsilon^T(\omega)$ be the truncation functional of $F_\varepsilon(\omega)$ as defined in (5.3). By Theorem 5.1, for $\mathbb{P}$-almost every $\omega \in X$ and for every sequence $\varepsilon_j \to 0$ there exists a subsequence (not relabelled) and a Carathéodory function

$f_0^T(\omega) : \Omega \times \mathbb{R}^{m \times d} \to [0, +\infty)$, which is quasiconvex in the second variable, such that for every $A \in \mathcal{A}^{\mathrm{reg}}(\Omega)$ and $u \in W^{1,p}(A; \mathbb{R}^m)$

$$\Gamma(L^p)\text{-}\lim_{j \to +\infty} F_{\varepsilon_j}^T(\omega)(u, A) = F^T(\omega)(u, A) := \int_A f_0^T(\omega)(x, \nabla u(x))dx.$$

Arguing as in the proof of Theorem 6.1 and using the characterization of non-homogeneous quasiconvex functions by their minima (see [2, Theorem II]), for every $M \in \mathbb{R}^{m \times d}$ and for almost every $x_0 \in \Omega$ we have

$f_0^T(\omega)(x_0, M)$

$$= \lim_{r \to 0} \frac{1}{r^d} \min \left\{ F^T(\omega)(u, Q_r(x_0)) : u - Mx \in W_0^{1,p}(Q_r(x_0); \mathbb{R}^m) \right\}$$

$$= \lim_{r \to 0} \lim_{j \to +\infty} \frac{1}{R_j^d} \inf \left\{ F_1^T(\omega)(v, R_j Q_1(r^{-1}x_0)) : v \in \mathcal{D}^{1,M}(R_j Q_1(r^{-1}x_0)) \right\}$$

$$= \lim_{r \to 0} \lim_{j \to +\infty} \frac{1}{R_j^d} \inf \left\{ F_1^T(\omega)(v, R_j Q_1(r^{-1}x_0)) : v \in \mathcal{D}^{T,M}(R_j Q_1(r^{-1}x_0)) \right\},$$

with $R_j = r/\varepsilon_j$. Set for any $A \in \mathcal{A}(\mathbb{R}^d)$

$$H^T(\omega)(M, A) := \inf \left\{ F_1^T(\omega)(v, A) : v \in \mathcal{D}^{T,M}(A) \right\}.$$

We now show that there exists $\phi_M^T : X \to [0, +\infty)$ such that for any $\bar{x} \in \mathbb{R}^d$

$$\lim_{R \to +\infty} \frac{1}{R^d} H^T(\omega)(M, R Q_1(\bar{x})) = \phi_M^T(\omega).$$

Set

$$\tilde{H}^T(\omega)(M, A) := \inf \left\{ \int_{B_T} \int_A f(\omega)(x, \xi, v(x + \xi) - v(x)) \, dx \, d\xi : v \in \mathcal{D}^{T,M}(A) \right\}.$$
$$(8.12)$$

Since

$$|\tilde{H}^T(\omega)(M, R Q_1(\bar{x})) - H^T(\omega)(M, R Q_1(\bar{x}))| \leq 2d C_1(\omega)(|M|^p + 1) T R^{d-1},$$

it suffices to show that

$$\lim_{R \to +\infty} \frac{1}{R^d} \tilde{H}^T(\omega)(M, R Q_1(\bar{x})) = \phi_M^T(\omega). \qquad (8.13)$$

For any $A \in \mathcal{V}$ denote $Q^A = \bigcup_{j \in A} Q_1(j)$ and $\Psi^T(M, A)(\omega) = \tilde{H}^T(\omega)(M, Q^A)$. Note that $\Psi^T(M, A) \in L^1(X, \mathbb{P})$ for any $A \in \mathcal{V}$ by conditions (8.2) and (8.4), and that by (8.12), Ψ^T is subadditive, according to Definition 8.1 (ii). Moreover, since f is statistically homogeneous, Ψ^T is stationary, according to Definition 8.1 (i). Condition (iii) of Definition 8.1 is trivially satisfied, since f takes values in $[0, +\infty)$. Hence, by applying Theorem 8.1 to the process Ψ^T we deduce that there exists $\phi_M^T \in L^1(X, \mathbb{P})$ such that for every $N > 0$

$$\lim_{R \to +\infty} \sup \left\{ \left| \frac{\tilde{H}^T(\omega)(M, RQ_1(x))}{R^d} - \phi_M^T(\omega) \right| : |x| \le N \right\} = 0,$$

which in particular yields (8.13). This on the one hand implies that $f_\omega^T(x_0, M)$ does not depend on the first variable and, on the other hand, that it does not depend on the subsequence $\{\varepsilon_j\}$. Hence, the whole family $F_\varepsilon^T(\omega)$ Γ-converges to $F^T(\omega)$. Arguing as in the proof of Theorem 6.1 we get the same result for $F_\varepsilon(\omega)$. If τ is ergodic, then $f^T(M)$ and F^T are deterministic, so is $f(M)$ and (8.11) holds.   □

We complete this section by providing a typical example of a random density $\mathtt{f} = \mathtt{f}(\omega, \xi, z)$.

Example 8.1 (A Random Density) Let $\mathcal{Y}$ be a Poisson point process in $\mathbb{R}^d$ of intensity 1; that is, $\mathcal{Y}$ is a point process such that for any bounded Borel set A in $\mathbb{R}^d$ the number of points of $\mathcal{Y}$ in A has a Poisson distribution with parameter $|A|$, and for any N and any disjoint bounded Borel sets $A_1, \ldots, A_N$ the random variables $\#(A_1 \cap \mathcal{Y}), \ldots, \#(A_N \cap \mathcal{Y})$ are independent.

Let φ be a $C_0^\infty(\mathbb{R}^d)$ function such that $\varphi \ge 0$. We set

$$f(\omega)(x, \xi, z) = \chi_{B_1}(\xi)(1 + |z|)^p + \left(1 + \sum_j \varphi(x - Y_j)\right)^{-1} \left(\sum_j \varphi(x - Y_j)\right)$$

$$\times h(\xi)(1 + |z|)^p,$$

where $\mathcal{Y}(\omega) = \bigcup_{j=1}^\infty Y_j(\omega)$, $p > 1$, and $h(\cdot)$ is a non-negative function that satisfies the following upper bound:

$$h(\xi) \le \frac{C}{(1 + |\xi|)^{d+p+\delta}}$$

for some $\delta > 0$. In this example $\mathtt{f}(\omega, \xi, z) = f(\omega)(0, \xi, z)$ and, by Remark 2.4, the growth assumptions are satisfied.

References

1. Braides, A., Piatnitski, A.: Homogenization of random convolution energies. J. Lond. Math Soc. (2) **104**, 295–319 (2021)
2. Dal Maso, G., Modica, L.: Integral functionals determined by their minima. Rend. Semin. Mat. Univ. Padova **76**, 255–267 (1986)
3. Krengel, U., Pyke, R.: Uniform pointwise ergodic theorems for classes of averaging sets and multiparameter subadditive processes. Stochastic Process. Appl. **26**, 289–296 (1987)

Chapter 9
Application to Convex Gradient Flows

Abstract So far, we dealt with static problems by studying the Γ-convergence of families of functionals $\{F_\varepsilon\}$. Here, by taking advantage of the Γ-convergence results proved, we analyze some dynamical aspects by considering gradient flows for a convex family $\{F_\varepsilon\}$, and prove the stability of the gradient flows with respect to Γ-convergence by applying the minimizing movements scheme along this family of functionals.

Keywords Gradient flows of convex energies · Stability by Γ-convergence · Minimizing movements along sequences of functionals · Homogenized gradient flows · p-Laplacian evolution equation

9.1 The Minimizing-Movement Approach to Gradient Flows

We first recall some definitions and relevant results to our end (see [1] and [5, Chapter 8]). Let $F : L^2(\Omega; \mathbb{R}^m) \to [0, +\infty]$ be a weakly lower semicontinuous proper (that is, not identically equal to $+\infty$) functional. We introduce a time-scale parameter $\tau > 0$ and we solve the sequence of minimum problems starting from the initial value $u_0 \in L^2(\Omega; \mathbb{R}^m)$; that is,

$$\begin{cases} u_n^\tau \in \operatorname*{arg\,min}_{u \in L^2(\Omega;\mathbb{R}^m)} \left\{ F(u) + \frac{1}{2\tau} \|u - u_{n-1}^\tau\|_{L^2(\Omega;\mathbb{R}^m)}^2 \right\}, & n \geq 1 \\ u_0^\tau \equiv u_0 . \end{cases} \tag{9.1}$$

By applying the direct method of Calculus of Variations we get that the solutions u_n^τ exist for any $n \in \mathbb{N}$. The sequence $\{u_n^\tau\}_{n \geq 0}$ is called a *discrete solution* of the scheme (9.1) (implicit Euler scheme). A discrete solution extends to an interpolation curve $u^\tau : [0, +\infty) \to L^2(\Omega; \mathbb{R}^m)$ defined by

$$u^\tau(t) := u_n^\tau, \quad t \in [(n-1)\tau, n\tau). \tag{9.2}$$

Definition 9.1 (Minimizing Movements) A curve $u : [0, +\infty) \to L^2(\Omega; \mathbb{R}^m)$ is
called a *minimizing movement for F from* u_0 if u^τ, defined as in (9.1) and (9.2), up
to subsequences, converge to u as $\tau \to 0$ uniformly on compact sets of $[0, +\infty)$.

In this chapter we will deal with convex functionals F for which weak lower
semicontinuity is equivalent to strong lower semicontinuity. Moreover, by the
convexity of F, the functional $u \mapsto F(u) + c\|u - v\|^2_{L^2(\Omega;\mathbb{R}^m)}$ is strictly convex
for every fixed $c > 0$ and $v \in L^2(\Omega; \mathbb{R}^m)$; hence, the solutions u_n^τ are also unique
for any $n \in \mathbb{N}$. Then there exists a unique minimizing movement for F from u_0,
belonging to $C^{1/2}([0, +\infty); L^2(\Omega; \mathbb{R}^m))$ (see [5, Theorem 11.1]).

If $F(u) < +\infty$ then a *subgradient* of F at u is a function $\varphi \in L^2(\Omega; \mathbb{R}^m)$
satisfying the following inequality

$$F(v) \geq F(u) + \int_\Omega \langle \varphi(x), v(x) - u(x) \rangle dx, \tag{9.3}$$

for every $v \in L^2(\Omega; \mathbb{R}^m)$. For each $u \in L^2(\Omega; \mathbb{R}^m)$ we denote by $\partial F(u)$ the set
of all subgradients of F at u. The *subdifferential* of F is the multivalued mapping
∂F which assigns the set $\partial F(u)$ to each u. The domain of ∂F is given by the set
$\mathrm{dom}\, \partial F = \{u \in L^2(\Omega; \mathbb{R}^m) \mid \partial F(u) \neq \emptyset\}$. If $u \in \mathrm{dom}\, \partial F$ then there exists a unique
element of $\partial F(u)$ having minimal norm that is denoted by

$$\partial^0 F(u) := \underset{\varphi \in \partial F(u)}{\arg\min} \|\varphi\|_{L^2(\Omega;\mathbb{R}^m)}, \tag{9.4}$$

see for instance [1, Section 1.4].

Definition 9.2 ([1, Definition 1.3.2, Remark 1.3.3]) A locally absolutely continuous map $u : [0, +\infty) \to L^2(\Omega; \mathbb{R}^m)$ is called a *curve of maximal slope for F* if it
satisfies the energy identity

$$F(u(t_1)) - F(u(t_2)) = \frac{1}{2} \int_{t_1}^{t_2} \|u'(s)\|^2_{L^2(\Omega;\mathbb{R}^m)} ds + \frac{1}{2} \int_{t_1}^{t_2} \|\partial^0 F(u(s))\|^2_{L^2(\Omega;\mathbb{R}^m)} ds, \tag{9.5}$$

for any interval $[t_1, t_2] \subset [0, +\infty)$.

Note that u, as in Definition 9.2, satisfies (9.5) if and only if $u \in W^{1,2}_{loc}([0, +\infty);$
$L^2(\Omega; \mathbb{R}^m))$ and it is solution to the *gradient flow* equation

$$u'(t) = -\partial^0 F(u(t)), \quad \text{for almost every } t > 0, \tag{9.6}$$

see for instance [1, Corollary 1.4.2].

Lemma 9.1 *Let* $F : L^2(\Omega; \mathbb{R}^m) \to [0, +\infty]$ *be a proper, convex and lower-*
semicontinuous functional. Then, for every $u_0 \in L^2(\Omega; \mathbb{R}^m)$, *with* $F(u_0) < +\infty,$

there exists a unique minimizing movement $u \in W_{loc}^{1,2}([0, +\infty); L^2(\Omega; \mathbb{R}^m))$ for F from u_0 which is the unique solution to the gradient flow (9.6) with initial condition $u(0) = u_0$.

Proof We already observed that for such functional F there exists a unique minimizing movement $u \in C^{1/2}([0, +\infty); L^2(\Omega; \mathbb{R}^m))$. By Ambrosio et al. [1, Theorem 2.3.3], we have that the minimizing movement is a curve of maximal slope, which concludes the proof. □

Instead of a single functional, we now consider a family of functionals $F_\varepsilon : L^2(\Omega; \mathbb{R}^m) \to [0, +\infty]$ for $\varepsilon > 0$, that are proper, lower semicontinuous and convex, and $u_0^\varepsilon \in L^2(\Omega; \mathbb{R}^m)$ a given family of initial data. We apply the minimizing-movement scheme (9.1) with F_ε in place of F and, similarly, we get that there exists a unique discrete solution $\{u_n^{\tau,\varepsilon}\}_{n\geq 0}$ and an interpolation curve $u^{\tau,\varepsilon} : [0, +\infty) \to L^2(\Omega; \mathbb{R}^m)$ as in (9.1) and (9.2), respectively, depending on the parameter ε.

Definition 9.3 (Minimizing Movements Along Families of Functionals) Consider $u_0^\varepsilon \to u_0$ in $L^2(\Omega; \mathbb{R}^m)$ and let $\{\tau_\varepsilon\}_{\varepsilon>0}$ be a family of positive parameters such that $\tau_\varepsilon \to 0$ as $\varepsilon \to 0$. A curve $u : [0, +\infty) \to L^2(\Omega; \mathbb{R}^m)$ is called a *minimizing movement along* $\{F_\varepsilon\}_{\varepsilon>0}$ *from* u_0^ε *at rate* τ_ε if, up to subsequences, $u^{\tau_\varepsilon, \varepsilon}$ converges to u, as $\varepsilon \to 0$, on compact subsets of $[0, +\infty)$.

Note that, in general, the minimizing movements u may depend on the rate τ_ε (see e.g. [5, Example 8.2] and [4]). This does not occur for convex families of functionals, for which gradient flows are stable with respect to Γ-convergence, as stated in the following result (see [5, Theorem 11.2]).

Theorem 9.1 (Stability of Minimizing Movements Along Convex Functionals) *Let H be a Hilbert space and let $\phi_\varepsilon : H \to [0, +\infty]$ be a equi-coercive family of convex functionals Γ-converging to ϕ and let $x_0^\varepsilon \to x_0$ be such that $\sup_\varepsilon \phi_\varepsilon(x_0^\varepsilon) < +\infty$. Suppose that for every $\varepsilon > 0$ there exists a minimizing movement of ϕ_ε from x_0^ε. Then*

(i) *the family of minimizing movements of ϕ_ε from x_0^ε converge to the minimizing movement of ϕ from x_0;*
(ii) *for every rate τ_ε the minimizing movement along the sequence F_ε from x_0^ε coincides with the same minimizing movement of ϕ from x_0.*

The following theorem states the stability of the Γ-limit of convolution-type energies with respect to gradient flows; for technical reasons, we assume $p \geq 2$.

Theorem 9.2 (Convergence of Gradient Flows) *Let $p \geq 2$ and let F_ε be defined as in (2.1) with f_ε convex in the last variable. Assume that (H0)–(H2) hold and that F_ε $\Gamma(L^p)$-converge to $F : L^p(\Omega; \mathbb{R}^m) \to [0, +\infty]$, as $\varepsilon \to 0$. Let $\{u_0^\varepsilon\} \subset L^p(\Omega; \mathbb{R}^m)$ be a given family of initial data such that*

$$\sup_{\varepsilon>0} F_\varepsilon(u_0^\varepsilon) < +\infty, \quad u_0^\varepsilon \to u_0 \text{ in } L^2(\Omega; \mathbb{R}^m) \tag{9.7}$$

as $\varepsilon \to 0$. Let $\bar{u}$ and u^ε be the minimizing movements for F from u_0 and for F_ε from u_0^ε, respectively. Then, for every $\tau_\varepsilon \to 0$ as $\varepsilon \to 0$, we have that $\bar{u} : [0, +\infty) \to W^{1,p}(\Omega; \mathbb{R}^m)$ is the unique minimizing movement along $\{F_\varepsilon\}_{\varepsilon>0}$ from u_0^ε at rate τ_ε and satisfies

$$\bar{u} \in C^0([0, +\infty); L^p(\Omega; \mathbb{R}^m)) \cap W_{\mathrm{loc}}^{1,2}([0, +\infty); L^2(\Omega; \mathbb{R}^m)). \tag{9.8}$$

Moreover, we have that u^ε are solutions to the gradient flows for F_ε; i.e.,

$$\begin{cases} (u^\varepsilon)'(t) = -\partial^0 F_\varepsilon(u^\varepsilon(t)) & \text{for almost every } t > 0 \\ u^\varepsilon(0) = u_0^\varepsilon, \end{cases}$$

$\bar{u}$ is solution to the gradient flow for F; i.e.,

$$\begin{cases} u'(t) = -\partial^0 F(u(t)) & \text{for almost every } t > 0 \\ u(0) = u_0 \end{cases}$$

and $\{u^\varepsilon\}$ converges to $\bar{u}$ as follows

$$\lim_{\varepsilon \to 0} u^\varepsilon(t) = \bar{u}(t), \quad \text{in } L^p(\Omega; \mathbb{R}^m) \text{ for every } t > 0, \tag{9.9}$$

$$\lim_{\varepsilon \to 0} u^\varepsilon = \bar{u}, \quad \text{weakly in } W_{\mathrm{loc}}^{1,2}([0, +\infty); L^2(\Omega; \mathbb{R}^m)). \tag{9.10}$$

Proof We extend the functionals F_ε and F to $L^2(\Omega; \mathbb{R}^m)$ by setting $F_\varepsilon(u) = +\infty$ and $F(u) = +\infty$ for every $u \in L^2(\Omega; \mathbb{R}^m) \setminus L^p(\Omega; \mathbb{R}^m)$. Note that, by the assumptions on f_ε we have that the functionals F_ε are convex and lower semicontinuous in $L^2(\Omega; \mathbb{R}^m)$ and the same holds for F since it is a Γ-limit. By Corollary 4.2 every $\{u_\varepsilon\}_{\varepsilon>0} \subset L^2(\Omega; \mathbb{R}^m)$ satisfying

$$\sup_{\varepsilon>0}(\|u_\varepsilon\|_{L^2(\Omega;\mathbb{R}^m)} + F_\varepsilon(u_\varepsilon)) < +\infty \tag{9.11}$$

is precompact in $L^p(\Omega; \mathbb{R}^m)$ and therefore in $L^2(\Omega; \mathbb{R}^m)$. Moreover, every limit is in $W^{1,p}(\Omega; \mathbb{R}^m)$. Hence, we get that $\{F_\varepsilon\}$ is L^2-equicoercive and $\Gamma(L^2)$-converges to F, as $\varepsilon \to 0$. Note that at fixed $\varepsilon > 0$ the functionals F_ε are lower semicontinuous and coercive with respect to the weak L^2-topology, which is sufficient to deduce the existence of the corresponding minimizing movements. We can then apply Theorem 9.1. Thus, there exists a unique minimizing movement along $\{F_\varepsilon\}$ from u_0^ε at rate τ_ε which coincides with $\bar{u}$, and $\{u^\varepsilon\}$ converge uniformly to $\bar{u}$ on compact subset of $[0, +\infty)$. By Lemma 9.1 we have that $\bar{u} \in W_{\mathrm{loc}}^{1,2}([0, +\infty); L^2(\Omega, \mathbb{R}^m))$ and $\bar{u}(t) \in W^{1,p}(\Omega; \mathbb{R}^m)$ for every $t \in [0, +\infty)$. Moreover, the decreasing behavior of F_ε along u^ε, (9.5) and (9.7) give (9.10). It remains to prove (9.9) and the continuity of $\bar{u}$ with respect to the L^p-topology. By the monotonicity of $F(\bar{u}(t))$ and (3.18), we infer that

$$\|\nabla \bar{u}(t)\|_{L^p(\Omega;\mathbb{R}^m)} \leq C(F(u_0) + 1)$$

for every $t \geq 0$. Therefore, by the strong L^2-continuity of $\bar{u}$, we get that $\bar{u}(s) \to \bar{u}(t)$ weakly in $W^{1,p}(\Omega; \mathbb{R}^m)$ as $s \to t$, which, in particular, gives (9.8). Finally, since $\{u^\varepsilon(t)\}$ satisfies (9.11) for every $t > 0$, then (9.9) is implied by (9.10).   □

9.2   Homogenized Flows for Convex Energies

In this section we apply the results obtained in Theorem 9.2 to the periodic-homogenization case. We describe the homogenized gradient flow in (9.14) and (9.15) under Neumann and Dirichlet boundary conditions, respectively.

Theorem 9.3 (Homogenized Gradient Flows) *Let $p \geq 2$, let F_ε be defined as in (2.1) with f_ε convex and C^2 in the last variable and let (6.1)–(6.6) hold. Let $\{u_0^\varepsilon\} \subset L^p(\Omega; \mathbb{R}^m)$ be a given family of initial data satisfying (9.7). Then,*

$$\partial^0 F_\varepsilon(u)(x) = \frac{1}{\varepsilon^{d+1}} \int_\Omega \nabla_z f\left(\frac{y}{\varepsilon}, \frac{x-y}{\varepsilon}, \frac{u(x)-u(y)}{\varepsilon}\right) dy$$
$$- \frac{1}{\varepsilon^{d+1}} \int_\Omega \nabla_z f\left(\frac{x}{\varepsilon}, \frac{y-x}{\varepsilon}, \frac{u(y)-u(x)}{\varepsilon}\right) dy, \qquad (9.12)$$

Theorem 9.2 holds and the family of solutions $\{u_\varepsilon\}$ to the gradient flows

$$\begin{cases} \partial_t u^\varepsilon = -\partial^0 F_\varepsilon(u), & in\ (0, +\infty) \times \Omega, \\ u^\varepsilon(0) = u_0^\varepsilon, \end{cases} \qquad (9.13)$$

converges weakly in $W_{loc}^{1,2}([0, +\infty); L^2(\Omega; \mathbb{R}^m))$ to the solution to the gradient flow

$$\begin{cases} \partial_t u = \mathrm{Div}(\nabla f_{hom}(\nabla u)), & in\ (0, +\infty) \times \Omega, \\ \nabla f_{hom}(\nabla u)\, \nu = 0, & on\ \partial\Omega, \\ u(0) = u_0, \end{cases} \qquad (9.14)$$

with f_{hom} defined by (6.18).

Proof By Theorem 6.2 we have that

$$\Gamma(L^p)\text{-}\lim_{\varepsilon\to 0} F_\varepsilon(u) = F(u) := \begin{cases} \int_\Omega f_{\text{hom}}(\nabla u(x))dx & u \in W^{1,p}(\Omega;\mathbb{R}^m) \\ +\infty & \text{otherwise}. \end{cases}$$

For every $u, v \in L^p(\Omega;\mathbb{R}^m)$, arguing as in the proof of Proposition 5.2, the first variation of F_ε is given by

$$\int_\Omega \left\langle \frac{\delta F_\varepsilon}{\delta u}(x), v(x) \right\rangle dx$$

$$= \frac{1}{\varepsilon^d} \int_\Omega \int_\Omega \left\langle \nabla_z f\left(\frac{x}{\varepsilon}, \frac{y-x}{\varepsilon}, \frac{u(y)-u(x)}{\varepsilon}\right), \frac{v(y)-v(x)}{\varepsilon} \right\rangle dy\, dx.$$

Thus (9.12) follows by the symmetric roles of x and y. For the limit functional F we have that

$$\text{dom}\, \partial F = \Big\{ u \in W^{1,p}(\Omega;\mathbb{R}^m) : \text{Div}(\nabla f_{\text{hom}}(\nabla u)) \in L^2(\Omega;\mathbb{R}^m),$$

$$\nabla f_{\text{hom}}(\nabla u)\nu = 0 \text{ on } \partial\Omega \Big\}$$

and, for every $u \in \text{dom}\, \partial F$, $\partial F(u)$ is single-valued and there holds

$$\partial^0 F(u) = -\text{Div}(\nabla f_{\text{hom}}(\nabla u)).$$

Hence the thesis follows by applying Theorem 9.2.   □

Reasoning as in the proof of Theorem 9.3, by Proposition 5.3 we obtain an analogue result for the homogenized gradient flow with Dirichlet boundary conditions. Note that, the L^p-equicoerciveness of the family $\{F_\varepsilon^{r,g}\}_{\varepsilon>0}$, defined in (5.20) and satisfying (H0), has been already shown in the proof of Proposition 5.4.

Theorem 9.4 *For fixed $p \geq 2$, $g \in W^{1,p}_{\text{loc}}(\mathbb{R}^d;\mathbb{R}^m)$, and $r > 0$, let $\mathcal{D}^{r\varepsilon,g}(\Omega)$ be defined by (5.19). Let F_ε and $\{u_0^\varepsilon\} \subset \mathcal{D}^{r\varepsilon,g}(\Omega)$ satisfy the hypotheses of Theorem 9.3 and let $F_\varepsilon^{r,g}$ be defined by (5.20). Then, Theorem 9.2 holds and the family of solutions $\{u_\varepsilon\}$ to the gradient flows*

$$\begin{cases} \partial_t u^\varepsilon = -\partial^0 F_\varepsilon(u^\varepsilon), & \text{in } (0,+\infty) \times \Omega, \\ u^\varepsilon = g, & \text{in } \{x \in \Omega \mid \text{dist}(x,\Omega^c) < r\varepsilon\}, \\ u^\varepsilon(0) = u_0^\varepsilon, \end{cases}$$

with $\partial^0 F_\varepsilon$ defined by (9.12), converges weakly in $W^{1,2}_{loc}([0, +\infty); L^2(\Omega; \mathbb{R}^m))$ to the solution to the gradient flow

$$\begin{cases} \partial_t u = \mathrm{Div}(\nabla f_{\mathrm{hom}}(\nabla u)), & \text{in } (0, +\infty) \times \Omega, \\ u = g, & \text{on } \partial\Omega, \\ u(0) = u_0, \end{cases} \tag{9.15}$$

where f_{hom} is defined by (6.18).

An interesting application of Theorem 9.3 is provided by the following example, which is then made more specific in Remark 9.1. For similar results related to p-Laplacian evolutions see also [2, 3].

Example 9.1 (Evolution of Purely-Convolution Operators) Let $a \in C^\infty_c(B_1)$ be a non-negative function. We consider functionals F_ε as in the statement of Theorem 9.3 with $f(y, \xi, z) = a(\xi)|z|^p/p$. For every $u \in \mathrm{dom}\,\partial F_\varepsilon$, formula (9.12) reads as

$$\partial^0 F_\varepsilon(u) = -\frac{1}{\varepsilon^{d+p}} \int_\Omega \left(a\left(\frac{x-y}{\varepsilon}\right) + a\left(\frac{y-x}{\varepsilon}\right) \right) |u(y) - u(x)|^{p-2}(u(y) - u(x))\, dy$$

and, by Theorem 9.3, the family of solutions $\{u_\varepsilon\}$ to the gradient flows for F_ε converges, as in (9.10), to the solution $\bar u$ to the gradient flow for the functional

$$F(u) := \begin{cases} \dfrac{1}{p} \int_{B_1} a(\xi) \int_\Omega |\nabla u(x)\xi|^p dx\, d\xi & u \in W^{1,p}(\Omega; \mathbb{R}^m) \\ +\infty & \text{otherwise,} \end{cases}$$

that is given by the following system of equations

$$\begin{cases} \partial_t u = \mathrm{Div}\left(\displaystyle\int_{B_1} a(\xi)\, |\nabla u\,\xi|^{p-2} (Du\,\xi \otimes \xi)\, d\xi \right) & \text{in } (0, +\infty) \times \Omega \\ \displaystyle\int_{B_1} a(\xi)\, |\nabla u\,\xi|^{p-2} (\nabla u\,\xi \otimes \xi)\, v\, d\xi = 0 & \text{on } \partial\Omega \\ u(0) = u_0. \end{cases} \tag{9.16}$$

We now consider the case $p = 2$ and $m = 1$. By Example 6.1, the Γ-limit is given by the quadratic form

$$F(u) := \begin{cases} \dfrac{1}{2} \int_\Omega \langle A_{\mathrm{hom}} \nabla u(x), \nabla u(x) \rangle dx & u \in W^{1,2}(\Omega) \\ +\infty & \text{otherwise} \end{cases}$$

where A_{hom} satisfies

$$(A_{\text{hom}})_{i,j} = \int_{B_1} a(\xi)\xi_i\xi_j d\xi, \quad \text{for every } 1 \leq i, j \leq d.$$

Hence, the homogenized gradient flow takes the form

$$\begin{cases} \partial_t u = \text{div}(A_{\text{hom}}\nabla u) & \text{in } (0, +\infty) \times \Omega \\ \langle A_{\text{hom}}\nabla u, \nu \rangle = 0 & \text{on } \partial\Omega \\ u(0) = u_0. \end{cases}$$

Remark 9.1 (Approximation of p-Laplacian Evolution Equation) In the scalar case $m = 1$, if we assume the kernel a to be also radially symmetric then $F(u) = c_p \int_\Omega |\nabla u(x)|^p dx$ where

$$c_p := \int_{B_1} a(\xi)|\xi_1|^p d\xi,$$

as shown in Example 3.1. Thus, formula (9.16) reduces to

$$\begin{cases} \partial_t u = c_p \Delta_p u & \text{in } (0, +\infty) \times \Omega \\ \langle \nabla u, \nu \rangle = 0 & \text{on } \partial\Omega \\ u(0) = u_0. \end{cases}$$

References

1. Ambrosio, L., Gigli, N., Savaré, G.: Gradient Flows in Metric Spaces and in the Space of Probability Measures, 2nd edn. Lectures Math. ETH Zürich. Birkhäuser Verlag, Basel (2008)
2. Andreu, F., Mazón, J.M., Rossi, J.D., Toledo, J.: A nonlocal *p*-Laplacian evolution equation with Neumann boundary conditions. J. Math. Pures Appl. (9) **90**, 201–227 (2008)
3. Andreu, F., Mazón, J.M., Rossi, J.D., Toledo, J.: A nonlocal *p*-Laplacian evolution equation with nonhomogeneous Dirichlet boundary conditions. SIAM J. Math. Anal. **40**, 1815–1851 (2009)
4. Ansini, N., Braides, A., Zimmer, J.: Minimizing movements for oscillating energies: the critical regime. Proc. R. Soc. Edinb. Sect. A **149**, 719–737 (2019)
5. Braides, A.: Local Minimization, Variational Evolution and Γ-convergence, volume 2094 of Lecture Notes in Math. Springer, Cham (2014)

Index

© The Author(s), under exclusive license to Springer Nature Singapore Pte Ltd. 2023   115
R. Alicandro et al., *A Variational Theory of Convolution-Type Functionals*,
SpringerBriefs on PDEs and Data Science,
https://doi.org/10.1007/978-981-99-0685-7

Printed in the United States
by Baker & Taylor Publisher Services